CHASING LAKES

CHASING LAKES

Love, Science, and the Secrets of the Arctic

KATEY WALTER ANTHONY

HarperOne
An Imprint of HarperCollinsPublishers

HarperCollins books may be purchased for educational, business, or sales promotional use. For information, please email the Special Markets Department at SPsales@harpercollins.com.

FIRST EDITION

Designed by Bonni Leon-Berman
Illustrations by Ina Timling
Map by Katey and Peter Anthony
Unless otherwise noted, all photographs courtesy of the author

Library of Congress Cataloging-in-Publication Data has been applied for.

ISBN 978-0-06-300199-2

22 23 24 25 26 LSC 10 9 8 7 6 5 4 3 2 1

*For Mom, who showed me that flowers
can spring from ashes
And for Dad, who first taught me to inquire*

Contents

Prologue

There's a pond I've watched for the last twenty years. It sits behind the University of Alaska Fairbanks, where I'm a professor and scientist. That's what I do professionally: watch lakes and ponds, and share their secrets. It's not an ordinary pond like the type I grew up with in the high alpine meadows of the Sierra Nevada. This pond is a greedy pond, a gluttonous pond. This pond echoes the mystery that I've spent my professional life solving, as well as the deeper mysteries of my own heart that have been revealed over the years.

Not content to remain within the constraints of the topography around it, this pond actively lowers the landscape—day by day, year by year—carving for itself an ever-larger basin to fill with its own black water. Most students passing this pond along the bike path on their way to campus see only the sparkle on the water's surface and the ripples from the lesser scaup diving ducks, which make the pond their summer home. They don't notice this pond's tactics. But I'm watching this pond. I know its hunger.

The shore where I used to tie my research raft, back when I first started measuring the pond, was once underlain by ice-rich permafrost. The perennially frozen ground was covered with soft green hummocks of sphagnum moss and wild cranberries. A sparse canopy of spruce and paper birch trees had at one time shaded my rubber boat from this high-latitude region's midnight sun.

But heat from the pond has been melting the ground ice. As the pond's margin collapsed, berries were drowned. Now only a few dead tree trunks stand above the water's surface. Around the pond, other trees are tipping, falling, and disappearing into the pond, as the very earth in which they are rooted gives way to the pond's appetite.

I watch as a mating pair of lesser scaup glide between flooded snags. In their wake, I see a sprinkle of silvery bubbles rise to the black water's surface. The dime-size bubbles dance out, coming to rest in a perfect, two-foot-wide circle, where they pause for a moment before bursting to join our atmosphere.

On this breezeless day, I see more spurts of bubbles rise across the pond, radiating out into bubble-circles of different sizes that linger for only a second before they, too, pop. I know these bubbles well. They are the burping and belching of methane gas released from the pond's murky digestive tract, where microbes living in the cold, dark sediments feast on the remains of ancient plants and animals buried beneath thick, sticky mud. These bubbles hold a secret.

With funnel-like homemade bubble traps I invented as a graduate student while studying lakes in Siberia, I've captured bubbles on this pond, bottled them up, and analyzed their methane content in a laboratory for decades. What I've learned from the bubbles I've collected on this pond, and many others like it, is that a unique source of methane dominates the bubbles from these Arctic lakes, particularly lakes beneath which permafrost is rapidly thawing.

Ancient carbon, locked away as the remains of plants and animals that died and became incorporated into permafrost soils thousands of years ago, is being released as bubbles of methane. Compared to an equal amount of carbon dioxide, methane is a far more potent greenhouse gas, contributing to climate warming by trapping the sun's radiation and heating Earth's surface.

Once the permafrost thaw process has begun, this pond and millions of other permafrost thaw lakes strewn across the Arctic will continue to consume the frozen soils around them, generating more methane, which in turn leads to more warming and thawing. This vicious feedback cycle threatens to accelerate climate warming beyond what most models predict.

I arrived at this pond in 2000, as a twenty-four-year-old graduate student who thought that survival in this world depended on my own hard work and an iron will. I moved away from my family at the age of twelve. In 1992, at age sixteen, I went to live on my own for a year in Russia in the aftermath of the fall of the Soviet Union. Having discovered as a child the serenity and escape from humanity that remote lakes offer, I came to Alaska in 2000 with a will to be among such lakes, to know them and to learn the secrets of their scientific importance.

Today, I am the wife of a Minnesota farmer who splits his time between seeking ways to grow food more sustainably in the US Midwest and assisting me in the Arctic with science. I am also the mother and joint homeschool teacher of our two boys, who often accompany us on our research trips into the wilderness. Coming to terms with my tripartite role of wife, mother, and scientist, I've had to face an entirely new kind of challenge: the dark inner workings of my own heart. Like the university's pond, whose gluttony consumes the very shores that outline its existence, I, like all people of the human race, was born apt to take from people and my environment more than was my nature to give back. But there is hope for people as they age and mature, if their ears are open to wisdom and their hearts to change.

As a scientist and a Christian, I believe that if we are to be good stewards of this Earth we call home, we must start by knowing it. Knowing it requires time spent observing and watching, connecting pieces of information in our minds, and discerning what is right. We can propagate goodness as we share discoveries of the natural world with our children, grandparents, and friends by spending time with them in nature, studying it, drawing it, and writing about it.

As a scientist who wasn't convinced twenty years ago that climate change was immediately observable, and as a Christian who twenty years ago wasn't sure I believed in God, today I believe in both. We have a responsibility to take care of the planet and the people who live on it. We cannot afford to procrastinate any longer. We also have a choice about how we view change. We can adapt to it and look for ways to help make the changes positive. I'm thankful for the God-given hope that good will prevail. I hold my husband's hand, teach my boys, and return to work every morning with that hope.

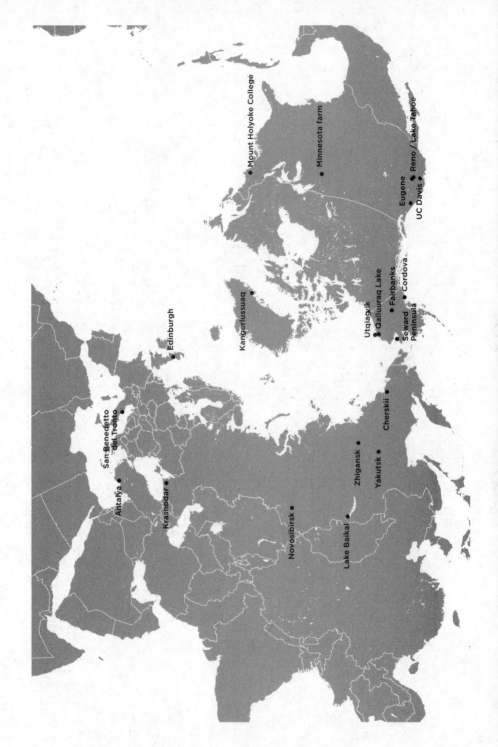

Polar map view of locations in my story

Part I

THE BUBBLE TROUBLE

ONE

An American Girl in Cherskii

K atya, I am beating your face on the asphalt," Sergey said to me in Russian.

I sat hugging my knees on the doorstep of Sergey's tattered yellow laboratory building, feeling like a lost child. The year was 2001, and I was at the end of my rope—a lowly twenty-five-year-old graduate student with seemingly no hope of accomplishing what I was meant to do.

By using this Russian expression, Sergey was taking a hard stand against my latest effort to get results. I'd suggested that instead of measuring the methane bubbles in the lake, which I wasn't doing well, I try to measure the methane coming out of plants growing in the Arctic lake-side wetlands.

"That is a waste of time." Sergey frowned at me. "Katya, if you don't follow my direction, you can spend your remaining time at my science station working as a cook instead of doing research."

These words stung. I gulped to keep my emotions hidden inside. My head burned with rage. I hadn't come all this way to be a cook. I was bound and determined to become a scientist. Since I'd left home as an adolescent, no one had ever threatened me about how

I spent my time. Time was mine. My own will was the force that steered the rudder of my life.

Sergey towered over me. The skin under his eyes fell in bags, forming deep creases above his cheeks. His lips pursed between a bushy mustache and an overgrown beard that concealed the lower half of his wizened face. He sighed and shrugged his shoulders, as if to ask himself what he was going to do with this foolish American girl hugging her knees outside his laboratory.

It wasn't just that I was a mess. I was an expensive mess. With the money it cost for me, an American graduate student, to be at the Northeast Science Station for six months, Sergey could have afforded to pay nine Russian students. For him, this meant I needed to produce what nine students collectively would have produced in my stead. For me, I needed to make something of myself, to do something important in science during my PhD opportunity, so that I could rise out of poverty and anonymity into the security and meaningfulness I longed for.

But I was failing—and failing miserably.

I'd come to Siberia to study methane emissions from Siberian lakes. I was attempting to go beyond the pioneering work of Sergey Zimov and Terry Chapin, my PhD advisors (see center insert photos 4 and 8). Four years before my arrival to Cherskii, Sergey and Terry had published a pivotal paper revealing that microbes living in the bottoms of Siberian lakes were capable of converting previously frozen Ice Age carbon from thawing permafrost soils into the potent greenhouse gas, methane.[1]

What made this more amazing was that they had proved the microbes were making methane at temperatures as low as 33°F, the temperature at which ice melts. Until Sergey and Terry's work, people had thought microbes in ecosystems shut down at such cold temperatures. But Sergey and Terry postulated that release of permafrost-derived methane from Siberian lakes in winter contributed significantly to the storage of methane in the atmosphere above the Arctic in winter. While methane is less abundant in the atmosphere than carbon dioxide, it has 84 times the global warming potential in a short (twenty-year) timescale.[2] The implication was that permafrost-derived methane could be an important lever in climate change.

The problem was, as is often the case in science, we didn't know what the actual rate was of methane emissions from lakes. We had only a rough estimate, based on a handful of measurements Sergey had taken. My job was to more carefully quantify how much methane was escaping from North Siberian permafrost thaw lakes. I needed desperately to measure the methane, not only for science, but also to prove to myself that I could do it.

As Sergey pivoted to leave, I turned my eyes eastward, away from him. I was trying to focus on anything but my failure. My gaze followed the undulating dirt road that ran along the top of the river bluff leading to the station. Several hundred yards down the road, an eight-foot-tall wooden church with a green roof and a tall conical steeple stood above the dwarf shrubs along the river's bluff. It was only a model building, not a real church. But why was it there? Certainly Sergey, a world-renowned scientist, hadn't built it, had he? Part of me wished the church wasn't there. It was a pretty little church, but it reminded me of a faith I'd decided to abandon nearly a decade earlier.

I turned my eyes back to the science station and watched Sergey strut home. Small clouds of red dust billowed around his calloused, bare heels, as they rose and fell on the backs of his intentionally crushed leather dress shoes. These self-made flip-flops were just one of the many manifestations of his slogan, "Be lazy, but smart." Not smart would be taking the time to properly adorn shoes needed only to walk short distances among the handful of buildings at his Northeast Science Station.

Northeast Science Station in Cherskii, Russia

Sergey disappeared into his house, and I sat still in the laboratory doorway. I was new to the Arctic, yet despite my research frustrations, for the first time in my life I felt at home. It wasn't just the cozy little science station comprised of four wooden houses and a simple laboratory building nestled together on an exposed bluff overlooking the vast tundra floodplain of the Panteleha and Kolyma Rivers (see center insert photo 5). I couldn't have asked for a more serene location as a base for my fieldwork. At times, as the only foreigner at the station, I had the yellow laboratory and one of the pretty wooden houses with big picture windows and a snug sleeping loft practically all to myself. When German, Italian, or other American scientists were present—usually never more than half a dozen at a time—I shared the house with them. I didn't mind that flushing the toilet meant dumping a bucket of river water into the latrine, and that there was little or no means of connecting to the outside world. The Zimovs, Sergey and his wife, Galina, recent empty nesters, lived in the white main house overlooking the river. Next to them lived another middle-aged Russian couple, friends of the Zimovs from their university days in Vladivostok. These families, along with two others, had moved together by their own volition to the Kolyma, a former Soviet gulag territory, in the 1970s. They built the station and raised their children together on this bluff, all the while carrying out scientific measurements of forests, tundra, permafrost, and lakes from the uninsulated laboratory in summer and their warm, heated kitchens in winter. Fed up with Sergey's ultimatums, two of the families eventually evacuated, leaving one house to stand empty and the other, my house, to serve as a residence for foreign scientists. The living room at the front of my house served as the communal dining area, in which a hired cook served meals for everyone at the station. It was during these meals that I consulted with the Russians about scientific measurements, but mostly I was on my own to figure out how to carry out research on permafrost and greenhouse gases.

I was floundering in my approach. I had built a fleet of funnel-shaped traps intended to capture bubbles containing methane in the lakes. But for a reason I didn't understand, the bubbles seemed to move around my traps, avoiding them. I knew this PhD research was my opportunity to make something of myself, yet so far I had demonstrated only that I was useless. I'd known the work wouldn't be easy, but I hadn't expected my lack of results to be so thorough and soul crushing. I'd been here for three months and worked around the clock to collect notebooks full of data, but the results of my data collection were paltry. Conscious of the exact cost of having me in Siberia, Sergey wanted to teach me to work smart, not just hard. He didn't mince words, hence the beating-my-face-on-the-asphalt line. He wanted to break me down before building me back up to think and act wisely, not just swiftly.

I thought back to three months ago, when I had gawked out the window of the twin-engine turboprop Antonov An-24 during the four-hour flight from Yakutsk into Cherskii. My eyes desperately absorbed the shapes, sizes, and distributions of the cornucopia of lakes that spread out over the lowlands of the Russian Arctic. My heart pounded at the sheer number of these lakes, and I wondered about the relationship I was about to begin with them. Would I be able to measure their methane emissions? Would my time in Siberia be fruitful, and would I produce results worthy of a doctoral degree?

Still huddled on the doorstep of the old, yellow lab, I pulled my knees in closer to my chest and bit one side of my tongue to hold back my tears. I knew I had no choice but to keep trying to capture methane bubbles in the lakes even though I didn't yet see a way to succeed.

...

I FELL IN LOVE with lakes as a child in the 1980s, when my father and grandfather took me hiking high up in the eastern Sierra Nevada mountains near Lake Tahoe, the largest alpine lake in North America. Tahoe is where I first learned to swim long distances through deep, cold water, to navigate a sailboard, and to ask questions about why the clarity of the lake's water was diminishing over time. However, my love for lakes was born at much smaller lakes, lakes perched higher in the mountains. These small lakes were harder to reach, requiring a strenuous day's climb up through the Jeffrey and Ponderosa pines, along gushing snowmelt-fed creeks, and around twenty-foot-tall granite boulders that had been split in two by the annual freezing and thawing action of water collecting in their cracks. When we reached these high alpine lakes, not a soul could be seen or heard. Dad, Grandpa, and I had the majesty of that granite kingdom to ourselves, and the lakes sparkled like sapphire jewels before us.

Eager to know what the world was like outside suburban Nevada, I applied in 1991 to be an exchange student in Latvia, a Baltic state that had just regained independence from crumbling Soviet control. My application to Latvia was diverted to Russia, and so in 1992, at age sixteen, I was one of two Americans placed in southern Russia after the fall of the Soviet Union by the American Field Services, an international youth exchange program. I landed in Russia with my knowledge of the Russian language limited to the alphabet and a few words I'd memorized on the transatlantic flight.

In Krasnodar, where I was to live for a year, hardly anyone had ever met an American. I was continuously received with awe, appreciation, and royal hospitality, reactions I had never before received during my childhood of bouncing back and forth between divorced parents and living off charity and food stamps. But Russia was a broken country, struggling to regain its legs, and so even the public-housing-reliant and food-stamp-supported life I'd lived in the United States seemed rich at first, compared to the conditions of toiling Russians.

Desperately homesick and frightened, mostly about getting lost in a city in which I had no map, no means of language, and no access to a telephone, I found myself sequestered in the concrete courtyard of the Armenian family I had been placed with, longing for the freedom and beauty of the Sierra Nevada lakes. I'd seen pictures of Lake Baikal, so I knew Russia must also have some beautiful and wild lakes. I dreamed of one day going to Lake Baikal, to see the immense, pristine Siberian lake that rivaled Lake Tahoe in water clarity but was the largest single freshwater repository on Earth. When I left Russia a year later in 1993, fluent in the language and smitten with the people and their history, I vowed to return to study Lake Baikal, the deepest and oldest lake in the world.

Through a National Security Education Program scholarship, I returned to Russia as a nineteen-year-old in 1995, this time to

Novosibirsk in Siberia, where I spent a summer exploring future potential career pathways in medicine and the environment—medicine because I had heard that doctors received good salaries, and the environment because I loved being outdoors.

At the Research Institute of Traumatology and Orthopedics, I walked the hallways where people with severe spinal deformities from the far reaches of Russia were crowded together—some were in wheelchairs, but many just sat or lay on the floor, wearing expressions on their faces that suggested they had lost all hope. In the operating room, dressed in a medical gown, slippers, and mask, I passed implements to the nurses and surgeons operating on backs. I was startled when one man, lying facedown and anesthetized on the operating table with a curtain drawn across his shoulders to separate his head from his wide-open back, suddenly lifted his head, opened his eyes, and asked me for a glass of water in Russian. The nurse explained that sometimes this happened and the purpose of the curtain was to prevent the patient from looking at their own surgery. Russian medical wards were interesting, but I lacked the level of selfless compassion needed to care for the sick. Then there was a restlessness in me to be outdoors, which I did not think would be satisfied within the confines of most clinics.

It was during this summer studying, but not identifying with, medicine that I first realized my career would involve lakes. Later that summer I made good on a promise to myself, taking the train to Lake Baikal. Standing on the grassy headlands of the Angara River, the outlet of Lake Baikal, I wondered about the vast expanse that lay beneath the wavy surface. So many mysteries seemed to lurk and dance in that great body of water.

Where did all the water come from, and how long did it stay in the lake before it flowed into the river? What was it like thousands of feet below the lake's surface? What types of unique life could be found in this lake, and would such discoveries have any significance to humankind? Beyond the reach of my gaze, somewhere on the adjacent shore stood a giant, 750-hectare pulp-and-paper mill. The lake's water, needed in vast quantities for pulp production, was so pristine it required little treatment. But what was the water's fate after the toxic sludge was released back into the lake by this factory? How did the mill affect the lake's ecology? These were the first real scientific questions I began to ask myself, and they put me on my path to studying freshwater lakes.

...

THE MIDSUMMER 2001 rain fell softly on the metal roof above my second-floor loft bedroom at Sergey's science station. Lying on my four-inch-thick Soviet mattress on the linoleum floor, I curled up tighter, appreciating the weight and warmth of the heavy wool blanket inside the cotton duvet. My pillow was huge and square, like all the pillows I'd ever laid my head on in various beds across Russia. After a long, soggy day of fieldwork, it felt good to finally be dry. I stared at the wood ceiling that shielded me from the rain, and I recalled how it was I had ended up in Siberia studying methane emissions at Sergey's Northeast Science Station.

It started at the end of my master's degree program at the University of California, Davis. In the spring of 2000, I was interviewing with a new potential PhD advisor, Terry Chapin, a terrestrial ecologist based at the University of Alaska. As a graduate student, I'd been studying water quality at Lake Tahoe and Lake Baikal, but no one had ever mentioned "thermokarst lakes" to me until Terry asked in the phone interview if I'd be interested in studying them in North Siberia for a PhD project. I told him I'd give this some thought, and the first thing I did after hanging up the receiver was look up in a dictionary what "thermokarst" meant. It appeared to be related to the term "karst," a sinkhole-type feature commonly found in limestone-rich regions like Florida and China, where groundwater dissolves calcium carbonate rock. I was pretty sure that limestone did not play a major role in the Arctic landscape process Terry had mentioned. Rather, "thermo" implied heat, a sinkhole-forming force that causes the ice in permafrost ground to melt. Terry and Sergey had been working together for half a decade to begin revealing to the world the importance of permafrost in moderating the global carbon cycle and Earth's climate. They needed a graduate student to work on this project, and one who had knowledge of both lakes and the Russian language.

When I finally landed in Cherskii in April 2001, Sergey immediately took me out to see Shuchi Lake, one of the thousands of thermokarst lakes I had seen flying into Cherskii. Like other thermokarst lakes, Shuchi originally formed when a wedge of ice, larger than a van, melted in the permafrost ground. As the ice melted, the ground's surface collapsed, creating a sinkhole. The sinkhole filled with dark, tea-colored water and became a tiny pond. The dark pond water absorbed the summer sun's rays and further warmed the icy soils beneath. This warming caused more ground ice to melt, which in turn caused the pond to grow deeper and wider as the ground collapsed further. During the pond's initial years, the liquid water standing in this thermokarst depression froze completely in winter, pausing the permafrost-thaw process. But as the cycle of thaw and collapse progressed summer after summer, the pond grew too deep for the water to freeze to its muddy bed in winter. Soon enough, it became a full-blown thermokarst lake.[3]

When we first reached the lake's edge, Sergey pointed to an area of calm water in a moat where the winter ice was melting. "There is methane," he said. I looked and saw rising streams of bubbles break the water's surface before radiating out into circles of floating bubbles. A moment later, the bubbles popped and disappeared. Methane had entered the atmosphere.

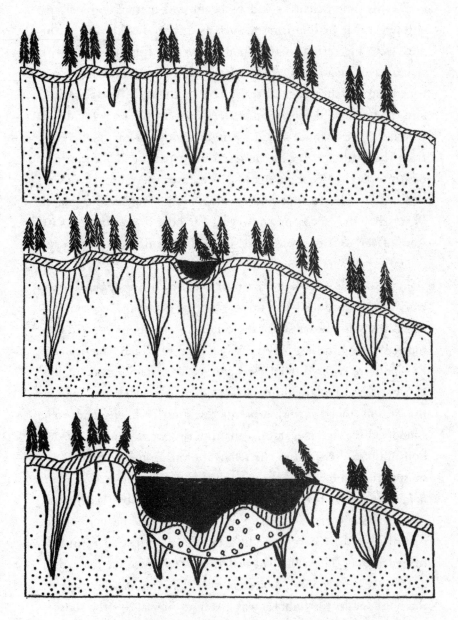

Thermokarst lake formation in cross section: *(top)* permafrost
soil with massive ice wedges; *(middle)* small thaw pond
forming above a melting ice wedge; and *(bottom)*
full-blown thermokarst lake

By this time, scientists widely recognized that amounts of anthropogenic greenhouse gases, such as carbon dioxide and methane, were increasing in the atmosphere at a frighteningly swift rate as a result of industrialization.[4] Researchers were also documenting the shrinkage of sea ice cover, the melting of glaciers, and the northward expansion of trees and shrubs. They pondered whether these changes were a natural response to Earth emerging from the Little Ice Age or the result of increasing concentrations of atmospheric greenhouse gases.[5] Whatever the cause, there was concern that irreversible planetwide changes were taking place. What they did not know, however, was how important the vast store of carbon frozen in Arctic permafrost soils would be to accelerating climate change.

In the decade preceding my arrival to Cherskii, Sergey and Terry's work had introduced the importance of Siberian "yedoma"— thick, icy permafrost soils (see center insert photo 7). Yedoma formed during the last Ice Age, when most of North America and Europe were covered by glaciers and ice sheets, and a vast region of Siberia had remained unglaciated. Fast sedimentation rates in this region, together with a productive grassland ecosystem that supported Ice Age megafauna (mammoths, woolly rhinos, bison, lions, horses, etc.), led to the removal of an immense pool of atmospheric carbon that became locked away in yedoma.[6] In their fieldwork near Cherskii, Sergey and Terry had observed higher carbon dioxide emissions from areas where yedoma permafrost was disturbed by natural or human causes compared to undisturbed areas.[7] They also showed that when yedoma thaws beneath lakes, methane bubbles emerge.[8] The radiocarbon age of the methane in the bubbles was a way to fingerprint the source. Methane in bubbles at Shuchi Lake had the same radiocarbon age as the plants and animals that had died during the last Ice Age. This meant that the ancient yedoma soil carbon was becoming

Remnants of the Ice Age ecosystem, previously frozen in
permafrost, thaw beneath thermokarst lakes.
This feeds microbes that produce methane.

food for microbes living in the oxygen-free lake bottom. It was
these microbes that were belching out bubbles of methane.

The evidence for the basis of my PhD research was mounting.
When I arrived in Cherskii, Sergey was working on calculations
in his living room that would eventually show that the amount of
yedoma soil carbon—which consists of frozen Ice Age plant and
animal remains—was equivalent to more than half the amount of
carbon in all of Earth's atmosphere.[9] Given the sheer magnitude of
this immense permafrost carbon reservoir, and Sergey and Terry's
first evidence that opening the permafrost freezer door leads to the
release of ancient carbon as greenhouse gases, it became apparent
that we needed to know more, before it was too late, before climate
warming progressed beyond our ability to try to close the freezer
door and keep the carbon in the ground. The goals of my PhD and

career research suddenly took shape: trying to get to the bottom of this process and its implications. Are thermokarst lakes a globally important funnel for releasing ancient methane into the modern atmosphere? How much of this permafrost soil carbon will thaw and be converted by microbes into greenhouse gases that will accelerate the warming of our planet? How soon will these changes take place? Was there any hope of stopping the greenhouse gas emissions? But, after fifteen weeks in Siberia, I was failing to answer the very first question of my career, "How do I trap a methane bubble?"

TWO

The Trapper

During the four years I called the Northeast Science Station home, I studied Shuchi Lake the most. There was a plethora of lakes in the Cherskii region, which had been a center for gulags during the Soviet era. Two of my study lakes, Blood Lake and Tube Dispenser Lake, had been named according to local legend. Following a 1973 uprising, eighty political prisoners were recaptured, lined up at the top of Blood Lake's steep bluff, and shot. While no one had been buried in Tube Dispenser Lake (see center insert photo 1), an old medical tuberculosis dispensary had been built upon its shores, and it was safe to assume it was the resting ground for a lot of plastic junk. In addition to lacking any sordid history, Shuchi possessed two major benefits. First, it was a good representation of thermokarst lakes formed in yedoma permafrost. This made it a perfect specimen to study. Second, Shuchi was convenient to get to—relatively.

Every morning at the Northeast Science Station, after washing up in the latrine and braiding my hair, I joined Sergey, the other Russian residents, and any visiting foreign scientists for breakfast. Half of the seats at the breakfast table doubled as couches or beds at other times of the day. Large picture windows provided a view of the laboratory slightly below us on the bluff above the long, twisting Panteleha River, Rodinka Mountain to the north, and the White Arrow Mountains on the eastern horizon. Here, Sergey was in his element. He would be already holding court at breakfast whenever I entered. He would sit upright on the sofa with his demure wife, Galina (Galya, as he called her), beside him. He had selected Galya for a wife because, as a student, she survived in a house that had been blown off the top of a mountain in Chukotka. He knew their future lives in the Arctic would require that kind of resilience. Entering the dining room, I noticed the stark contrast between the brown, leathery skin of Sergey's cheeks—a testament to decades of hot, dry summers and frigid, long winters—and the baby-soft white skin of his seemingly pudgy arms, arms that were almost always protected by clothing from mosquitos or cold.

Mealtime was an opportunity Sergey seized to inform the "bourgeois"—as he called us foreign scientists—about his philosophies on matters such as the healthfulness of smoking cigarettes and the purity of the Russian kitchen. "You see, Katya," he explained as I took my seat on this late July morning, "our nomadic ancestors living in yurts always kept a fire going. To be warm, even the children slept close to the fire. And so the lungs of northern people evolved over thousands of years to need some amount of smoke in them to flourish." Oblivious to his failure to convince me, he would ask if anyone would like to join him out on the wooden porch while he smoked. Upon his return, Sergey continued to philosophize, "The Russian kitchen uses very little spice, typically nothing more than salt, pepper, and garlic. The reason is that Russia is a cold country with plenty of means to preserve its meat. Spices are needed in warm countries where the meat more easily spoils."

I cast my eyes appreciatively across the breakfast table covered with porridge, hot tea, a large plate of crepes, and bowls of sour cream and jams made of cranberries or blueberries gathered the previous fall by Galya from the local tundra. After breakfast, I donned my mosquito jacket and Xtratuf rubber boots and ran down to the yellow lab. So full of excitement for all there was to do, I simply could not restrain myself to walk. I filled my large blue Kelty backpack with sampling gear that had been set out to dry the night before. I was ready to do my work.

Two weeks had passed since Sergey's ultimatum that I continue to study lakes if I didn't want to work as the cook. Submitting to him required mustering a meager amount of humility within myself, but this obedience also steered me toward further cementing my bond to him and to the lakes. This day I was eager to be out on the open water, measuring the lake's many properties.

Getting to Shuchi Lake was only a ten-minute walk north through the larch forest behind the station, but the thick trees and shrubs challenged my commute by snaring my homemade bubble traps and the oars strapped to my forty-pound backpack. Untangling the snares required setting down my pack and the cases containing instruments for measuring the lake water column's temperature, pH, salinity, oxygen, chlorophyll, and light penetration properties. I did not resent these inconveniences. I embraced them. Alone in the forest with the scent of Labrador tea wafting up from beneath my rubber boots and a cloud of mosquitos buzzing around my head, I knew my decision to pursue ecology rather than medicine had been the right one. I was at home now in my calling to solve scientific problems in the Arctic wilderness. I'd made it to the wilderness; now I needed to figure out how to solve some bubble-trapping problems. Eventually, I made my way down the steep eastern margin of the lake, navigating across cracks in the earth where the permafrost was actively thawing and giving way to puddles of liquefied mud that oozed toward the lake. There, finally, was my boat.

I was glad I didn't have to carry my watercraft to the lake. This had been a real worry before arriving, but luckily Sergey, with pride in his eyes, had presented me with an old, gutted aluminum motorboat on my first day. He had pulled it up from the bottom of the river. It wasn't a grand vessel, but it was mine, and it would do the job. The boat had two-inch-tall walls dividing the bottom of the ten-foot-long craft into three sections, like a TV dinner tray holding different dishes. I found this very useful for separating water-sensitive equipment from wet or muddy gear. My practice was to put my backpack—with field notebook, camera, extra batteries, sample bottles, labeling tape, and rain jacket—behind me to keep dry.

I lifted my boat from the shrubs and gave it a few pushes so that most of it was floating in the water before I loaded my gear and jumped in myself. Then, using one of the heavy oars, I shoved off, afloat on the black water. My oars squeaked in their locks, and their metal shafts clanked against the rim of the gutted motorboat with each revolving stroke as I rowed across Shuchi Lake. Once I had gotten accustomed to operating the heavy, mismatched oars, I found my boat quite maneuverable. I highly appreciated the level of independence and efficiency a limnologist (a person who studies freshwater ecosystems) has when she can carry her own field gear and get along working from an inflatable rubber raft or lightweight watercraft. I regularly told myself that this was an advantage over the often much better funded field of oceanography, in which ship time is expensive and logistically constrained. Anyway, it's all I had.

Science is the art of measurement. While quantifying this myriad of lake properties was tedious week after week, it gave me a four-dimensional understanding of a lake's character. My instruments allowed me to see over time how the surface and bottom waters slowly evolved as spring warmed into summer, and how a lake's mood would swing suddenly when cold air came into the region. Cold air caused this lake's surface water to cool in a very short time. The cold, dense water would sink, displacing the methane-rich bottom water, which

Shuchi Lake

rose to the surface to rage and vent for a few days until conditions
calmed down and a new peaceful equilibrium could be reached. I
knew that in lakes elsewhere in the world this seasonal turnover of
the water column could lead to a big pulse release of methane that
had been stored not as bubbles, but in the lake's bottom water itself.
Turnover was one more way that large quantities of methane could
escape lakes, and I needed to know how this compared to the bubbling in Siberia's thermokarst lakes.

Among my measurement tools—in addition to a Hydrolab, light
meter, and fluorometer, used for measuring the lake's chemical, physical, and biological profiles throughout the water column—were my
tightly coiled secchi disk and a Van Dorn sample bottle. My secchi
disk looked like an eight-inch dinner plate divided into alternating
quarters painted white and black. Lowering the disk into the water

with a rope, I could observe the exact depth at which the disk disappeared from view. This traditional approach, invented in the late 1800s, is still used today by limnologists to measure the transparency of lake water. The Van Dorn bottle, an elaborate eighteen-inch PVC pipe with a spring-loaded toilet plunger on each end, could also be lowered into the lake by rope. At a desired depth, I could send a nine-ounce bronze messenger down the line to trigger the spring release, causing the plungers to snap shut, encasing water from that unique depth inside the bottle. I could then haul the Van Dorn up to the lake's surface to pour its contents into various plastic and glass sample bottles for laboratory analyses of nutrients, dissolved gases, and organic carbon.

In the very front of the boat were my large and unwieldy bubble traps—my most important measurement tool and one of the few I'd built completely from scratch (see center insert photo 6). Lots of people had studied methane dissolved in the waters of North American and European lakes, but few people had studied lake methane bubbles, let alone in Siberian thermokarst lakes. When you're helping to pioneer a new topic, you need to pioneer the way to study it. I needed a method in which to capture the methane escaping from the thawing permafrost in the gut of Shuchi Lake: if I could trap the methane, I could measure the methane.

I thought back to my efforts in May, two months earlier. My first step to inventing my bubble traps had been to clean up the roadways.

My homemade methane bubble traps were made of an assortment of "trash" and "rubble." I had collected every empty plastic bottle I could find: these would trap the methane bubbles. With a huge supply of discarded 1.5-liter Sprite, Coke, and brown plastic bottles that once had held Russian beer, I returned to the station and cut two of the large blue tarps I had brought from Alaska into eight-foot circular shapes. I attached the mouth of each bottle to a small hole I had cut in the center of each tarp circle. Then I needed to maintain the shape of these funnel-like tarps while they were in the water. So I went to the Pile of Rubble. The Pile of Rubble sat on the rocky bluff just outside Sergey's west-facing kitchen window, where he could cast a watchful eye with ease. The Pile contained the remnants of previous expeditions and experiments that no one dared to throw away properly, in case they might come in handy later.

I was grateful to find coils of 8-gauge metal wire I could pull out and cut with strong pliers. I wove the wire in and out around the circumference of the circular tarps. After several days of struggling to get these large blue bubble traps to maintain a funnel shape when placed in the lake's water, I decided that my traps were too large and unwieldy. Besides, I was going to need a lot of bubble traps, and my supply of tarps was limited. Even then, in May, after only a month in Cherskii, I'd felt the pressure of my mission, and I'd wasted precious days on a dead-end solution.

I had asked Sergey if there was another material I could use to construct my bubble traps. He pointed out some plastic sheeting under the deck of my house, the type of sheeting locals typically used to grow vegetables inside makeshift greenhouses. The transparent plastic was perfect—thick enough to not easily tear but thin enough to be pierced by sharp wire without my needing to cut each little hole as I threaded the wire around the perimeter. Cinching the plastic and pulling up on the center of the circle produced my desired funnel-shaped plastic skirt that would hold its shape underwater.

I attached an inverted plastic bottle to the top of each skirt. On the bottles I'd installed a short one-inch-long plastic tube with a three-way stopcock. I used electrical tape and special marine Goop glue to seal the connections. It was important that when I submerged my traps in the lake, there were no holes or leaks through which the gas bubbles could escape.

The final touches involved tying bricks or pieces of scrap metal I had found in abandoned military outposts surrounding Cherskii to the bottom of each trap so it would be anchored to the lake bottom. To the top of each trap, I fastened another plastic bottle, closed with its own lid, to serve as a buoy and marker. Installing these traps at various locations around the lake entailed dropping an anchor line, tying a bubble trap to it, turning the trap upside down to let it fill with water, and then turning it right-side up again to hang submerged in the water column, filled completely with water. In working mode, the water in my trap would get displaced by bubbles percolating up from the lake bottom.

Now, sitting in my boat on a bright July morning, I looked out across the lake and thought about methane, the gas I'd come all this way to study. Unlike carbon dioxide, which has a high solubility (it dissolves in water, a property that makes it great for fizzy drinks), methane has a low solubility and forms bubbles in lake sediments, where it's produced by microbes decomposing dead organic matter. The bubbles erupt out of the sediments when their volume overwhelms that of the sediment pore spaces that hold them. If the water's surface is still when a person is gazing down at it, he or she will see a stream of silvery bubbles percolate up from the dark water's depths. The bubbles typically travel as small clusters, wobbling upward and taking the form of dime-size backward-moving parachutes. As the bubbles ascend, they swell, since their size is inversely proportional to the amount of water pressure above them. I could see spurts of bubbles breaking the surface at various locations around Shuchi Lake.

My hope and aim were that bubbles releasing from the lake-bed sediments would rise into my traps and be funneled up, displacing the water in the inverted bottles with the closed stopcocks at the top. The transparent plastic sheeting of the bubble-traps gave me the advantage of viewing any bubbles trapped on the plastic skirt under the surface of the lake. A few good shakes of the skirt would dislodge those bubbles and send them gliding up along the inner face of the skirt-funnel, into my plastic collection bottle.

With my fleet of twenty-five bubble traps set all around Shuchi Lake, I was ready to catch and measure the belching methane. At least I'd thought I was.

THREE

War of Attrition

At the science station, days of the week blended together. For me, there was no such thing as a weekend, only a self-imposed long list of research tasks. One Sunday morning, the wooden model church on the bluff, which I'd learned had been built by people from Cherskii as a destination place to walk three miles from town, caught my eye. I swung my backpack over my shoulder, exited the lab, and turned onto the forest footpath to Shuchi Lake. Since resolving at the age of seventeen to live as if there was no God, I had rarely stepped into any church. My mother, living her life far away in northern Nevada, was still a strong believer. I knew she prayed for me. When I spoke to her on the phone from California or Alaska, or hiked with her in the Sierra Nevada on visits, Bible verses she'd memorized would usher forth from her mouth. Sometimes I feigned that I wasn't listening. Other times I retorted that her faith didn't pertain to me. But all of this was a bluff to hide the cutting response the spoken scriptures were having on my heart. In truth, I was thirsty for what she had to say. My heart felt that she was speaking from streams of living water. I wanted her to tell me more, but my pride kept me from saying that. Alone now in Cherskii, God was a concept that was becoming harder to dismiss. When I'd asked Sergey what he thought about God, he just told me, "Between me and God, it is complicated." He said nothing more than that, but he didn't ridicule me for asking or deny that there was a God.

On this early-August Sunday morning, I rowed around a silent Shuchi Lake to check my array of bubble traps. At first, I excitedly pulled on the buoys and lifted the traps to the lake's surface so I could see if they had captured any bubbles. If they had, I would then take from my backpack a homemade graduated cylinder, a bottle with volumetrically calibrated tick marks made with a Sharpie. I would fill this special bottle with water and invert it just near the lake's surface. Carefully positioning the trap's stopcock beneath it, I would then open the valve. A stream of bubbles would rush out of the trap, through the open stopcock, and into my calibration bottle. By noting the date, time, and volume of gas trapped, I was able to calculate the flux rate of bubbling from the lake.

But too often the flux I recorded was zero. Zero bubbling. These zero fluxes measured by my traps baffled me since I could see with my own eyes as I rowed around that the lake was indeed bubbling. Sergey had said these lake bubbles were the crux of the threat of permafrost thaw contributing to global climate warming. But why, then, were the bubbles not going into my traps? After nearly ten weeks of paltry trapping, I was at a loss. I returned to the science station deflated and disappointed. I could see that the methane bubbles were out there, but why wasn't I catching them?

A week later, when I arrived at the lake, I noticed arcing lines of perforations and quarter-size holes in the plastic sheeting of some traps. On other traps, strings tying the trap to the anchor had been severed. On still others, the stopcocks were missing. My heart sank. I was already barely making progress, but now I was going backward. Completely befuddled, I brought one of my traps back to the station and showed Sergey. "Muskrats," he said. I was up against a formidable foe: a family of muskrats was staging a devastating and coordinated attack on my bubble traps.

Muskrats look like their cousins, beavers, only they are smaller and have scaly, ratlike tails almost as long as their bodies. Native to North America, they were introduced to several southeast Siberian river valleys in the early 1930s, and it's little wonder they found their way to northeast Siberia, where their need of large supplies of fresh water and vegetation are met handsomely.

My initial tactic against the guerrilla attacks was to replace the traps. But building new traps was no small task: it required ransacking more roadways and ramshackle buildings for supplies. And since I still had a battery of routine measurements to maintain on a circuit of lakes in the region, there was only one time when I could afford to make new traps: when I was supposed to be asleep.

For several nights, with the midnight sun streaming in through small windows, I sequestered myself in the attic of the yellow laboratory, building my traps amidst piles of boxes and science equipment. Each morning around 4 a.m., I dragged myself down to the lower level of the old building with my replacement traps, put on my rubber boots at the door, and picked my way across the silent grounds of the science station, where every other soul had long been asleep in their beds. I paused to look at early morning dew forming on the cotton grass growing in wet areas of the station grounds and then lifted my eyes to watch the fog slowly wind its way up the river below me. The world and sky glowed a brilliant pink and yellow, like a sunset. These colors as well as the dew and fog would all be gone when I woke at 9 a.m. for breakfast, but for now I relished them. Who knows, I thought. Maybe I was the only person in the whole far north of Russia to be taking in this wonderful landscape. It felt like a secret reward for my work. It was enough to make me forget, for a moment, my war with the muskrats. But only for a moment.

As quickly as I built my traps, the muskrats chewed through them. My defense needed to be stronger. My supply of stopcocks from the United States was limited. I needed extra deterrents. I began mixing hot red pepper from the kitchen into the glue I used on the stopcocks, hoping to ward off the water rats. This family of muskrats unfortunately possessed a preference for red pepper since the spicy traps were most frequently violated. By this time, I had constructed over sixty traps just to maintain my revolving fleet of twenty-five at Shuchi Lake. A month passed of this power struggle. I was losing the battle—and my sanity, it seemed—and I had no more supplies to waste.

News of my month-long war of attrition with the fiendish and wily muskrats soon spread around the station and dominated mealtime conversations. A wealthy museum owner from Moscow wanted to help. Sasha was visiting Cherskii to pay for mammoth ivory scavenged by jobless locals in the region. He was a middle-aged and pale-skinned man, slightly overweight with big glasses and a big smile. Sasha sang my name as he beckoned me toward him on the porch one morning.

"Katyaaaa, how would you like to have a muskrat hat?" he asked. "I will do you a favor. I will hunt the muskrat."

Sure enough, Sasha went out and shot a muskrat, later showing me the carcass in the shadows beneath my front porch. That evening after supper, Sasha cornered me in the kitchen. My back was turned and my hands were plunged in a sink of soapy water, washing dishes. I cringed as I heard my name sung out again— "*Katyaaaa*"—as Sasha's voice drew nearer. Suddenly he was right behind me. "Do you want to thank me?" He put his hands on my bottom, apparently thinking it was the reward he should have for shooting a muskrat. I turned and pushed past him, telling him I'd thank him in a more public location. I took a seat in the corner of the dining room, remaining silent among other people's post-dinner conversations. At the least, I hoped, my sexual harassment would correlate with a diminished fighting force on Team Muskrat.

The next day, I went out to see if I'd gained the upper hand with at least one of their fighting forces picked off. But, losing a family member seemed only to have solidified their resolve to foil my scientific work. The muskrat attack on my traps accelerated thereafter. In my seemingly endless effort to mend and build new traps ahead of their attacks, I became tired and weak. At times, I wondered if I was losing my mind.

The dread of my traps' destruction was in the forefront of my thoughts one morning. I had done everything within my human power to fight the muskrats. Not knowing where else to turn, I went upstairs after eating breakfast instead of down to the lab. I knelt on the linoleum floor next to the disheveled wool blanket on my mattress and put my forehead beneath my hands to pray silently, but heartily, to a God I did not fully trust in and whom I doubted was even real. I prayed that God might help me finish the job I had come here to do.

That day, when I went to Shuchi Lake, not a single trap was broken. There were no new teeth marks, no strings severed, no stopcocks lost.

The next day I went out again. No stopcocks lost, no strings severed, no new teeth marks. I did not see another sign that muskrats had been there the rest of the summer—nor during the four remaining years of my doctoral field research.

Finally, the war was over.

...

THE MUSKRAT PROBLEM may have been solved, but my failure to trap bubbles remained. Plus, it was now September, and my time would soon run out: I would be returning to Alaska for winter coursework at the university. What did I have to show for myself?

This was when I went to find Sergey.

When Sergey wasn't handling business or solving logistical challenges, he spent most of the summer in his underwear lying on his couch. He said his job was to think. He was a very important thinker. He said sometimes he thought so hard he started to sweat all over.

On his couch is where I expected to find him when I knocked and stuck my head through the mosquito netting that hung across the open doorway to announce my arrival. As always, Galya left her work in the kitchen to come greet me and invite me inside. Indeed, there Sergey lay. On some days, Sergey spoke to me with half-open eyes, not bothering to turn in my direction. In those moments I felt uncomfortable and unimportant. But on this day, he got up and went into the kitchen, where he lit an off-brand cigarette and hunkered down next to the woodstove chimney. Galya did not like smoking in the house, and she alone seemed to possess the power to influence Sergey's behavior. Keeping one outstretched hand on his cigarette in the woodstove, Sergey looked me square in the face.

His full attention was on me, and I did not take it for granted. This was my opportunity for advice, which I desperately needed. I took a deep breath and let my frustrations spill out. Despite my best efforts I was not catching bubbles in the lake. I re-announced my desire to expand my wetland measurements. I admitted to Sergey that I had placed box-type chambers over the sedges and equisetum plants that skirted the edges of the lakes. These plants were releasing methane to the atmosphere through tiny holes in their leafy tissues. Surely, if I changed gears and put my efforts toward quantifying plant-associated methane emissions I would at least succeed in capturing *some* methane.

Sergey didn't see it that way. "Methane from wetland plants is not new science. You will not make a discovery there." I could hear the disgust in his voice. "What people don't yet know about is yedoma carbon bubbling out of lakes. If you want to study wetland plants, you might as well stay in Alaska or California. What you need to discover here is the value and importance of yedoma permafrost. To do that you must study bubbles in the lake itself, not the wetland plants along the margins."

Sergey paused. He took a drag on his cigarette, exhaled into the woodstove, and then looked back at me. "Be smart in the use of your time in Cherskii."

Again he was thoughtful. "Didn't your grandmother teach you to pick up money the right way? If you come across bills scattered along the roadway, why should you spend your time picking up the one- and ten-dollar bills? Rather, you should focus on the one-hundred- and one-million-dollar bills."

I knew this was Sergey's way of telling me that I needed to learn to discern and pursue the rare but globally important scientific research questions and not waste time studying mundane topics that anybody could study anywhere.

Sergey stroked his bushy beard and smoked again. He said, "I am the big boat. You are the small boat. The big boat does not get off his couch very often because it maneuvers slowly, but from its great vantage point it can see long distances and can advise the little boat, which has much more mobility and speed, where to go and when."

Sergey advised that, in addition to the lakes, I start studying small ponds. I could compare the methane concentrations in the water column and bubbling dynamics of new little thermokarst ponds that had formed recently by melting yedoma ice wedges to that of human-made ponds dug in bedrock and sand deposits along the river's bluff. Since we didn't expect sand to produce methane, at least not anywhere near the levels of the thawing, organic-rich yedoma permafrost soils, then in this way I could show the importance of yedoma having elevated methane emissions.

I agreed with him. I wanted to comply and succeed, but my time that first summer would soon run out.

Sergey pierced me with his eyes. "Katya, I have been watching your character. People at my station see that you are a hard worker. They stretch their limits to pave a path for a girl who sells herself to science. You have an ambition they don't understand. Maybe you don't understand it yourself. But people see your potential. You wear a flag on your back that claims you can jump five meters. No one has ever jumped three meters. In the end, what will it be if you only jump four meters? It will be a complicated social situation. Just remember, there are nine students somewhere in the world who don't get to go into the field because of the money that NSF* has invested in your trip to Siberia."

That night I cowered in a ball and cried. I knew that I'd harbored unreal expectations for myself. Science was slow. Science was hard. And I was young and overeager. My future depended on my success, but I was failing.

Most of the next night Sergey and I spent over vodka and a bottle of wine at his kitchen table. Putting his glass down and reaching to smoke, Sergey said, "If I had ten Katyas to work with, two would pull through, two would jump to five meters."

"I hope I would be one of those two," I said in earnestness, but I was fearful of his response.

Sergey looked at me and nodded. "You *are* one," he said. "I have seen enough foreign scientists. You can make it to academia." Sergey put his cigarette out in the woodstove, latched its door, and returned to the table. The creases in his face revealed the passage

---

* National Science Foundation

of time. "No one at my station any longer has time to practice the science of lakes. Now the fruit of my work is the reputation of this station. The stakes are high on your head, Katya."

In his stern way, Sergey had given me confidence. That's what I needed.

I resolved to return the next year with rededicated effort. And the next year and the year after. Whatever it took. If I'd been able as a teenager to keep my promise to myself of one day seeing and studying Lake Baikal, then as a graduate student I would come back to Siberia and quantify the methane seeping into our atmosphere as permafrost thawed beneath these myriad lakes and ponds.

FOUR

Cats in the Ice

Arriving early at the shore of Shuchi Lake one early September day a year later, I paused before getting into my boat. Summer was again at its end in the Russian far north. The mosquitos were gone. Rays of yellow sun beamed down through parting clouds. How quickly things had changed in just a few days! Green shrubs and larch trees had transformed into a fireworks display reflecting off the lake's surface. The larch needles were a brilliant yellow. Beneath this golden canopy, purple and magenta herbs, and red, orange, and yellow willow and dwarf-birch shrubs were a backdrop behind the tall green sedges growing at the lake's edge.

This had been my second summer in Siberia. I had expanded my study to continuous, long-term monitoring of bubble traps in three large lakes and six smaller ponds. To that I had added short-term, single-episode measurements at twenty-one other lakes around the region. I had hired an American undergraduate student as a field assistant for the summer and had been given command of a small outboard motorboat and a van to reach lakes along the region's riverways and roads.

"You can sit on this bucket," Sergey had said to me when he

handed me the keys to the green Mitsubishi van he'd recently pro-
cured. Having been rolled by the previous owner, the van's body
was in bad shape and lacked a front windshield altogether, but
Sergey had gotten a good deal on it. I quickly got used to steering
the totaled Japanese vehicle from the right side. My driver's seat was
an upturned five-gallon bucket and I wore laboratory safety goggles
to shield my eyes from dust and rain. Despite my geographically
expanded effort, this summer had gone similarly to the previous
one: some success capturing bubbles, but overall mostly inconsis-
tent progress. It was frustrating, but not entirely disheartening like
the previous summer. I was twenty-six years old, two years into my
PhD program, and was coming to terms with the reality of data.
They may not reflect my hopes and ideals, but the more data I had
(and my stack of filled field notebooks was getting quite tall), the
more confident I was that my data reflected the truth in the natural
world around me.

In my pursuit of truth, I had also started attending the local
evangelical church in Cherskii on Sunday mornings. The bench in
the back was a nonthreatening environment where I could ponder
questions of faith. Besides, in the back, no one would notice when
a tear slipped down my cheek, from the well of nostalgia for hymns
I hadn't sung in the ten years since being an exchange student in
Krasnodar. Driving my totaled fieldwork van in town was illegal,
so Sergey let me drive his Land Rover to Cherskii on Sundays, provided I also did the weekly bread run for the station. What I hadn't
understood at the time was that getting to know local Cherskii residents at church would ultimately mean success in my data collection efforts too.

This early September day, I climbed into my boat and glided out
onto Shuchi's glassy water. Suddenly, the palette of autumn colors
reflected in the black water became blurred. First there were isolated

patches of ripples in the water's surface. Then, all around my boat, the entire lake started to bubble at once. With the arrival of fall, a low-pressure weather system was moving through Cherskii, causing the barometric pressure to drop. The lake was reacting. This sudden drop in pressure caused bubbles stored in sediments to expand, triggering tremendous ebullition in the lake.

With this much bubbling going on, surely my traps would have gas in them, I thought. My mind scurried faster than my oars could row. As I moved through the circuit of traps, my excitement faded to disappointment. Some of my traps had caught a lot of gas, but not in proportion to the abundant bubble streams I'd seen rising around me in the lake. I hauled the boat up onto shore and walked back to the station. How many days had I returned to the station disappointed that my efforts were apparently failing?

When the lake finally froze five weeks later and my assistant had returned to the US, I checked the thermometer that hung outside Sergey's house. The temperature had fallen to –10°F. I donned my wool sweater, my down jacket with its duct-tape patches, and the muskrat hat that had been presented to me after Sasha's hunting escapade, and I walked down the road from the station atop the Panteleha River bluff. Steam was rising from the river fifty feet below, but the water was flowing and not yet frozen solid. I knew the time had come to work indoors. The cold air hurt my lungs to breathe. The condensation of my breath froze and formed ice crystals on my eyelashes, but being indoors was not what I dreaded. Rather, before me was the daunting task of bringing back to life the Shimadzu gas chromatograph on the first floor of the yellow lab. This machine would help measure the exact amount of methane I had collected in my scant bubble-trap sample set.

My traps gave information on the *volume of gas* released from the lake, but they could not tell us *how much methane* was in the bubbles. I had been storing my samples in glass bottles in a dark box in the refrigerator at the back of the lab. Determining methane concentration required the analysis of a gas chromatograph (GC). The manual was in Japanese with brief, poorly constructed sentences of English translation. I realized that reviving this instrument would require me to revive the girl I had been in childhood, the only one in my mother's single-parent home who had been able to tweak the knobs, wires, and screws on the VCR to make it work. I hadn't needed to fix electrical equipment in nearly twenty years. If anything, I had succumbed to a mindset of intimidation on such matters. In this case, I had no choice but to overcome myself.

"The gas chromatograph is very simple," Sergey told me.

So simple that he was not willing to come down to the lab to help me get it going. Discussing the principles of gas chromatography with Sergey in his living room and reading and thinking between the lines of the broken Japanese translation, I figured out how to connect a tank of hydrogen and a separate nitrogen generator to the GC. With power and a secondary regulator on the GC, I could control the flow rate of nitrogen through the long coil of tubing inside the GC's oven. The hydrogen gas produced a small constant flame. When I injected a precise 5-milliliter volume of bubble-trapped gas through a port connecting to the nitrogen stream, the flow of nitrogen carried my gas sample through the long tubing until it reached the hydrogen flame. At this point of intersection, the hydrogen flame would leap higher since my bubble sample contained methane, a flammable natural gas. The strength of the flame was directly proportional to the concentration of methane in my sample. The flame's signal was recorded as a spike on graph paper, which I printed out and then manually entered into my laptop computer. I was proud of myself for pushing past intimidation to solve this technical hurdle.

Now I would be able to quantify the methane concentration in my bubble samples—and this was the start of a slow-burn scientific epiphany. I'd known that bubbles from other lakes being studied around the world had clocked in at around 20 to 60 percent methane. Samples I had collected in the human-made ponds not affected by yedoma thermokarst were within this same range. But as I started to run samples from bubble traps that had been placed along the margins of thermokarst lakes, places where yedoma permafrost was most rapidly thawing, I saw the concentrations of methane shoot up. One sample was 80 percent, another 85 percent. I couldn't believe my eyes—until I found samples with 90 percent and even 95 percent! These very high concentrations indicated that yedoma thermokarst methane bubbles were special.

But this was not my only discovery during that trip.

...

FIVE DAYS AFTER the lake froze, Sergey made an announcement at breakfast. "Katya, today I will go with you to Shuchi Lake." It was a startling statement. Sergey hadn't been to the lake with me since the first April when he had presented me with my metal boat. I thought this was very strange, especially since there was nothing we could do there now on this mid-October day when the lake was covered by thin ice. "We must go as soon as we finish our breakfast," Sergey insisted.

Rather than taking my trail through the woods, we drove the longer circuit around the bluff to reach the lake. This, I thought, exemplified Sergey's philosophy to be lazy, yet smart. He had frequently told me that he did not believe in "sports exercises," which is what he called the physical exertion he witnessed in most of my fieldwork.

On the drive, Sergey continued, "It is imperative to go to the lake today. The ice cover is thick enough to hold us and it has not snowed yet."

We slid out of his Land Rover and clamored down the bluff to the lake's edge. Sergey carried a six-foot-long iron crowbar, which I soon learned was an ice spear. I had been out on lake ice once or twice before on winter camping trips with friends at snow-covered lakes in the Sierra Nevada, but never before had I stood at the edge of a perfectly smooth, black, ice-covered lake with the notion of stepping forward.

I took a few small steps out and looked straight down through the ice, which appeared thinner than my forearm. I could see the lake bottom from which brown and green mosses protruded. Little black water beetles swam under the ice beneath my feet. Looking out farther, the muddy lake bottom plummeted out of sight where the water deepened. I assured myself that if the ice broke where I was, at least I could stand on the sediments, which were only about three feet under me.

Sergey must have sensed my fear because he said to me, "Don't worry, Katya. Autumn ice is friendly. It is strong and it makes cracking sounds to warn you before it breaks."

Sergey was much heavier than I was and had already walked thirty feet onto the lake before turning around to wave me on. I followed.

The ice did make cracking noises. Loud, sharp cracking sounds. The piercing sounds sent fearful bolts of pain across my stomach, causing me to bend forward as if to hold on to something that wasn't really there to save me. Finally, I came to trust Sergey's words that we were not going to perish by walking on thin ice. My legs became surer of themselves and I was able to focus on the real reason Sergey had brought me out to the lake this day.

I had not imagined when I woke up that morning that this would become the greatest aha moment of my scientific life. As we carefully picked our way across the lake, I could see why my traps had rarely captured a representative amount of the bubbling I knew to be taking place. I had placed my traps randomly across the lake, but 99 percent of the ice on the lake's surface was perfectly clear and black with no bubbles.

My traps had all happened to be placed in the areas of black ice.

However, just as the stars are scattered across the night sky, so were the thousands of bright, white bubble clusters trapped and frozen by the black winter ice sheet. While most of the ice sheet was black, I saw that at isolated points, the bubbles were clustered and stacked like white coins, one on top of another, in the ice sheet. I was completely elated to finally understand my problem. It was as if a veil had been lifted from my eyes and I now clearly saw why I had failed for so long to capture the bubbles. Given the rarity of the bubble patches, it was no wonder my traps had not overlapped with these source points of bubbling. The probability of placing a trap on one of these bubbling points was highly unlikely. It would be like shooting an arrow at the sky blindfolded and hitting a star instead of dark nothingness.

I gawked in wonder. The ice sheet was an incredible time-lapse photograph of bubbling. Not only did the ice layers reveal the frequency of bubbling but also, as a whole, the ice cover showed the patterns of bubbles. In some locations, individual dime-size bubbles clustered and stacked tightly together but did not touch (see center insert photo 2). In other clusters, stronger bubbling had caused bubbles trapped under the ice to bleed together. Eventually these pockets of gas had frozen in place like stacks of pancakes or even grand wedding cakes. Scientists had known for decades that bubbling in lakes was hard to quantify due to the spatial and temporal patchiness of bubbling, but never before had a scientist described what we saw in the ice this day. Here before us was a frozen map of the bubbling dynamic. A map that proved the key to a cornucopia of scientific discovery.

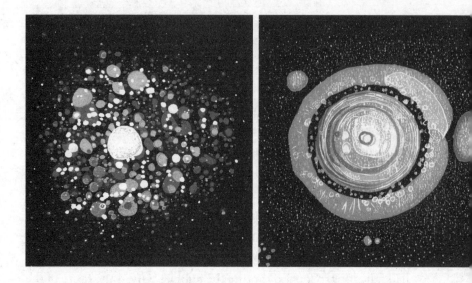

Why hadn't Sergey told me about this before, I wondered. Perhaps he thought I had to toil to earn appreciation for this gem of knowledge. Whatever the reason, Sergey was not the only person who knew about ice-trapped bubbles. A. N. Tolstoy, the great early twentieth-century Russian author and distant relative of Leo, had also known about them. In his Russian Christmas story, he wrote about children ice-skating on frozen lakes and pausing to light methane bubbles trapped in lake ice to heat their tea. This day Sergey Zimov had in mind the same activity. He planned to demonstrate the bubble's methane content without a gas chromatograph.

Sergey handed me the heavy iron ice spear and pulled a box of matches from his pocket. "You will stab the spear into that ice-bubble pocket. But—" he switched to English to emphasize a warning. Gesturing to the wild curly blond hair that encircled his head and beard, he said, "Be careful, Katya. Once I burned my whole vegetation."

I leaned back and then fell forward with the weight of the spear. It dashed through the ice and into the gas pocket. A great rushing sound ensued. Sergey, already on his knees with a lit match,

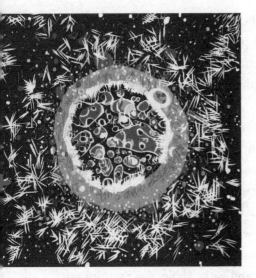

Methane bubbles trapped in
October lake ice

reached his hand out to ignite the wind of gas rushing out of the ice hole.

A flame burst upward, soaring above our heads as a great ball of fire. The heat from the flame together with the shock of what had just happened sent me hurling backward. When I stood again on my feet, I sensed an awful smell. I slipped off my gloves and touched my eyebrows. They were singed. Then I saw the charred remains of the curls at the bottom of my braid.

Sergey passed the matchbox to me and stepped aside to smoke as I commenced lighting more bubbles. My heart was pounding with excitement as I rushed from one bubble cluster to another, no doubt like one of the children in the Russian Christmas story. Each new bubble patch seemed to promise a better flame, a higher fire. I shouted excitedly, navigating this way and that among the thousands of bubble clusters, stabbing into their white pillows of trapped gas and setting alight as quickly as I could the torrent of gas that rushed out of the holes.

My shouts of surprise and awe enveloped the fireballs, some of which rose as high as the larch trees surrounding the lake (see center insert photo 3). When my excitement finally settled and I remembered that I was training to be a professional scientist, not one of Tolstoy's storybook children, I joined Sergey to take a long, careful look around the lake and absorb the wondrous revelations that the lake ice had to offer.

I noticed very few bubbles in the center of the lake. But as I approached the thermokarst margins, the density of bubble clusters increased hundreds of times. In addition to ice-trapped bubbles, there were manhole-size openings in the ice. Peering down through these holes, I saw regular streams of bubbles rising through the water column. These bubbles entrained warmer water from the bottom of the lake, generating convection that kept the ice holes open. These particularly strong seeps were found only along the thermokarst margins, where the permafrost was so actively and deeply thawing that it supplied a feast of ancient carbon to microbes generating the methane gas.

The Russian expression for thermokarst expansion literally translates as "lake eating into land." It's a highly accurate translation when one considers that the collapse and erosion of the thawed permafrost soils leads to digestion of organic matter by lake-bottom microbes. Accompanying that digestion, the microbes were now burping out methane manifested in the bubbles Sergey and I were witnessing.

That night, in his kitchen, Sergey and I decided that the various ice-trapped bubbles must be classified and systematically studied. Though Russians have known for a long time about their lakes' bubbles, because they are accustomed to living near lots of frozen lakes, no one had ever studied the makeup of the bubbles. Moreover, no one had ever classified the kinds of bubbles according to their shapes and bubbling rates. I figured this was because most scientists had put away their summer field gear and gone indoors to teach classes at universities in winter. I happened to be the lowly graduate student who had lingered outside a bit longer one year.

Sergey lit a cigarette and inhaled. His eyes, partially hidden by half-closed eyelids, moved their gaze toward the top of the wood-stove as he thought. He leaned over, cracked open the door to the chimney, and exhaled. When he returned to the table, Galya, who had been frying on the stove behind me, set a plate heaped high with dripping, hot, oily doughnuts in front of us next to other dishes of sardines, pickles, and fresh herbs from her greenhouse.

"The common Russian word for ice-trapped methane bubbles is *koshka*," Sergey explained.

I recognized *koshka* as the affectionate word frequently used for "cat." Calling the bubbles *koshka* likely reflected the resemblance of the bubble cluster patterns to a cat's paw print.

"We must exploit the richness of the Russian language," Sergey insisted. "We have many expressions for 'cat.'"

As we feasted on fish and doughnuts, we named three bubble-seep types according to their distinct ice-bubble patterns. *Kotenok* (type A), the weakest of the seep types, translates as "kitten." It refers to the individual bubbles that stack on top of one another in the winter ice sheet without merging laterally. *Koshka* (type B), named after a "tomcat," are the individual bubbles that laterally merge into larger bubbles under the ice prior to freezing. The third class, the largest-size bubbles, resembling a pancake-stack pattern, we called *kotara* (type C), after "a big, fat grandfather cat that sits on the woodstove." We identified a fourth ebullition seep class as a "hotspot," when there's steady enough bubbling to prevent ice from forming.

Having settled on our classification system, Sergey lit another cigarette and poured a round of vodka. He raised his glass and toasted, "To love." We soon all retired to our beds, but I was too excited to sleep. I knew I'd had my breakthrough. With Sergey's help, I had invented a new method for quantifying methane emissions, one that would go on to teach the scientific community that bubbling in thermokarst lakes was not completely random. It occurred at specific spots that were easily identifiable if a person took advantage of the ice sheet as a map of bubbling. This also meant that I needed to do research in winter. What I did not know at this time, however, was that this new method would ultimately reveal that bubbling is the dominant way methane escapes from North Siberian lakes, and these bubbles are the basis of a climate-warming feedback loop linked to permafrost thaw.

...

I'D FINALLY HAD my breakthrough, but I could not leave Cherskii without putting my precious newfound knowledge to use. My previously "random" placement of traps would not do. To really understand how much methane bubbled out of lakes, I had to place traps in strategic locations. This way I could monitor the behavior of the discrete bubbling seeps that generated these ice-bubble patterns. Using the ice spear, I opened the lake ice and placed my traps beneath the surface where the bubble clusters occurred. Having broken a good pile of sturdy larch tree branches, I hung each trap from a branch that extended across the holes.

Now, for the first time in the two years I'd worked on the lakes, my traps were finally collecting bubbles. Loads of bubbles! I could stand and watch the bottles begin to fill right before my eyes in traps placed over the strongest bubbling seeps. Other traps, placed over the weaker type-A and type-B seeps, took one or more days to accumulate enough bubbles to sample. I had everything in place to gather data on how much methane was coming up through the lake: my traps and the categorization method that Sergey and I had created. The final component of my breakthrough was finding someone to continue to record the measurements while I was away for the winter so that I would have a full year's data set.

I knew villagers would gladly accept the payment I was offering to do scientific fieldwork in winter, but I settled on asking Dima Draluk, a thirty-year-old local Cherskii resident, former Buddhist, and present-day wild-man firefighter who walked the tundra months at a time when he was not working. He was also an avid bodybuilder and the bald backup preacher at the evangelical church I attended on Sunday mornings. Dima had helped me with fieldwork on my study lakes before. He did not own a motorized vehicle, and the walk from his apartment in Cherskii to Shuchi Lake, which took thirty minutes in summer, would be at least an hour in winter through the darkness, snow, and ice. We both knew that work outside in winter was not trivial. The days would darken. The temperatures would plummet. The ice would thicken. The only way to check these traps was to carefully open the ice each day with the spear, and by spring, there would be six-foot-tall piles of ice chips next to each trap.

Half laughing at himself, Dima agreed to help me. "You should be grateful I am the person you asked to help. Any other fool in the town would gladly take your money and fake the data. I will be honest," he said, and I trusted him. With Dima's invaluable help, I would be able to collect long-term (up to 359 days) data allowing me to determine that the bubbling rates were different and distinguishable among the four seep classes.

On a cold day in late November, Sergey loaded my bags, samples, and customs paperwork into the Land Rover and drove me to the Cherskii airport. Sergey was quiet as he drove. He had given me all the instruction he could for the time being, and now it was time for me to go to Alaska to analyze my data, take courses at the university, and prepare to come back the following spring.

My own heart was swelling with all kinds of feelings as I sat in the passenger seat, watching the townspeople shuffle down the gloomy roads, bundled in long winter coats and tall fur hats. I was relieved and elated to have finally made a breakthrough in understanding methane emissions from lakes, but at the same time I was sad to leave this man and place that had come to occupy a very precious spot in my heart. I hoped that in Alaska I would be able to turn my data into results of a caliber worth the displacement of those nine Russian students by one American girl in Cherskii.

FIVE

Northern Lights

Three days of travel back around the world to get from Cher-skii to Fairbanks, Alaska, gave me time to reflect on the importance of what had happened and to write out a long list of what I wanted to accomplish in Alaska. But first I would need to meet with Terry Chapin, my PhD advisor at the University of Alaska Fairbanks.

Terry had recently been elected as a member of the National Academy of Sciences and had authored eight hundred articles and books on terrestrial ecology and ecosystem gas exchange. He was well esteemed by everyone. I myself felt intimidated, doubting I was cut out for his merit of science. While his spirit was much gentler than Sergey's and his words were always full of encouragement, I still did not want to let him down.

Upon my return, I was invited by Terry to have dinner with him and his wife, Mimi. We would make pelmeni, traditional Russian dumplings, and talk about how the field season had gone. Terry and Mimi were at least ten years older than Sergey. Since Mimi was fluent in Russian and since Terry found Sergey's ideas to be novel and important, working with Sergey in Russia was a project Terry and Mimi had been eager to share together. They had visited Sergey and Galya in Cherskii on multiple occasions in the 1990s and taken Sergey under their wing, hosting him and his young family for part of a year when the Chapins were at UC Berkeley. Mimi taught Sergey some English, and together they helped Sergey translate his ideas for the broadly read scientific journals. Mimi repeatedly had tried to impress upon Sergey the virtue of and need for humility, and I could tell she discerned from my stories that he still had a long way to go. By the end of that evening, Terry and I agreed to meet on campus soon to go over some of the data sets I'd collected in Siberia.

November days are short in the north, and the sun had long set by the time I left Terry and Mimi's house. I could see the northern lights, brilliant curtains of green and red dancing across the sky as I drove the thirty-minute stretch between Terry and Mimi's house, on a hill overlooking Goldstream Valley, to my little cabin, built on top of a different set of hills known as Chena Ridge, southwest of Fairbanks. Many people prefer to live at higher elevations in interior Alaska to escape the temperature inversions that occur. In winter, minus-40-degree air and ice fog settle in the valley bottoms, while up in the hills, hundreds of feet higher, the air is often clear, sunny, and 10 to 20 degrees warmer.

Before ascending the 1,500-foot hill I lived on, I pulled my little black two-seater Toyota pickup truck into the Water Wagon, a water-filling station at the bottom of Chena Ridge. Here I could fill up four blue five-gallon jugs with the water I would need for drinking and washing all the next week at home. Like many students in Fairbanks, I lived in a dry cabin, which meant the cabin didn't have any water supply or plumbing. My bathroom was an outhouse, which I loved because it got me outside in the fresh air more frequently than I would have gone had I lived in a house with plumbing.

Only once did I make the mistake during winter of leaving my coat inside to run to the outhouse. On that occasion a moose had

Ice fog over the Tanana River, Alaska, 800 feet below my cabin

wandered into my yard and was nibbling the willow bushes protruding from the snow in the clearing of the paper birch forest between my outhouse and cabin. Since I had recently heard a report that a student in Anchorage had been killed, kicked to death by a moose on campus, I hadn't dared try to pick my way back to my cabin while the moose was in my yard. I stood hugging myself and trembling, mostly from the cold that was rapidly sapping all the heat from my body, and wondering how long it would take for the moose to get its fill and move on.

Parking my truck full of water jugs among shrubs on the slope beneath my cabin—my parking spot—I was acutely aware of my loneliness. I lived in the Fairbanks North Star Borough surrounded by ninety thousand people, but ironically I always felt much more alone there than I ever did out on the tundra of Siberia. Besides, my friends were all getting married, and I was still single. I had dated plenty of people, but my selfish motives for companionship and other conveniences lacked virtue. Heeding the warnings of my parents not to marry young as they did, I had never pursued marriage in those relationships. Their advice rang loud and clear in my head: "Get your education first, Katey. Don't make the same mistakes we did." It contradicted the advice of most Russians, who urged me more intensely with each passing year to pay a little less attention to science and more attention to my biological clock. "You will miss your opportunity to be a woman," the Russians at Sergey's station had warned me on many occasions. "A real woman is a wife and a mother. You will regret it if you let this pass you by," they expressed with deep concern. But I dismissed the Russian advice, telling myself it was just a cultural difference. On this evening, however, as I pulled up to my cabin alone, I wondered. Family. What did that mean, anyway? To me it had meant only brokenness. I had long ago learned to live alone, and yet there was a wound in my heart that the bandage of science could cover by distraction for only so long.

I opened the back hatch of my truck and hauled the water jugs in pairs up the steep snowy path to my cabin, hoping I wouldn't slip. Fortunately, embers from the fire I'd lit earlier in the day still burned in the woodstove and the cabin was warm. I tossed another log onto the coals and was grateful for its crackling sound, which provided at least some comfort.

My sixteen-by-twenty-foot gray plywood cabin was not much to look at from the outside, but inside I had made it pretty. Next to the woodstove, I'd hung a section of an old door my grandpa Sid had painted blue, a piece of modern art. Opposite it was a round wooden table with a purple cloth covering and my cello, whose richly lacquered wood gave an added feeling of warmth to the room. An eight-by-four-foot alcove in the northwest corner of the cabin served as my kitchen. Large picture windows on three walls in the

Alaska birch forest

main room framed views of the paper birch forest to the southeast, the Alaska Range to the southwest, and the spruce-forested Cripple Creek valley to the northwest.

During inversions, fog filled the valley, so I could look down from my cabin on a sea of clouds, the surface of which reflected the warm winter alpenglow light. My favorite place in the cabin was the tiny eight-by-eight-foot loft directly above the kitchen and overlooking the main room. The loft had windows cut near the floor that allowed heat from the woodstove up and provided a view of the blue painting and the dining table. The loft also doubled as my library and home office. There I sat, my small wooden

desk facing south, in the direction of Denali, the highest mountain peak in North America, and I analyzed the data I had collected in Siberia.

My secluded work routine was to jam-pack the woodstove with
logs to get the cabin temperature up to 80 degrees, and then I
crunched all my data in a bikini. If I wasn't able to take a tropical
vacation, I'd treat myself to a sauna staycation. Once all the data
was organized, I prepared graphs that I would bring into my meeting with Terry.

Several weeks later, after I had done my best to distill the data
I'd collected in Siberia into a few key graphs, I entered the Arctic
Health Research Building on the University of Alaska's West Ridge
campus. I made my way down the hallway and to a door labeled
TERRY CHAPIN. I knocked on the door already cracked open, and
Terry swiveled in his chair to face me.

"Katey! It's good to see you." Terry smiled, his blue eyes twin-
kling beneath raised eyebrows. Hair was thinning on Terry's head,
but his gray beard was full and rich. Though he was not tall, he was
strong in body and mind.

Terry was the greatest servant-leader I had ever met. It didn't
seem to matter to him that he was world-famous, a recipient of the
Kempe Award for Distinguished Ecologists, a former Guggenheim
Fellow, a participant in the Intergovernmental Panel on Climate
Change, and an author of hundreds of well-esteemed scientific articles and books. In groups, Terry always waited until everyone else
was served before he took a plate of food. He was the first to wash
dishes and spend his time helping others with even the most mi-
nuscule tasks. At a summer field camp he led annually in Alaska,
students woke each morning to the sound of Terry's fiddle outside
their tents. All of these characteristics were so unlike me, a person
who was always competing for first place. What I never quite got

used to, though, was Terry's silence. He had a way of staring at me, without blinking, for a long time, intent on listening to and pondering whatever I might have to say. I knew this was how he listened to everybody. Still, feeling terribly self-conscious, I stumbled over my words to fill the silence.

That day in his office, I first explained to him my struggle to capture methane on the Cherskii lakes, how my initial random placement of bubble traps had failed, and how finally, once the lakes had frozen, I was able to see that the bubbles were not random but they occurred in discrete locations.

Terry stared at me without blinking. Uncomfortable and aware that I should probably not use too much of his time, I sped up. "If I placed my traps directly over these point sources of bubbling," I explained, "I captured tremendous amounts of gas. In fact, some traps captured so much gas that if I didn't come to vent the gas within two hours, the traps would rise out of the lake and tip over."

Suddenly, Terry stopped me. "Wow!" he said, breaking that calm silence that so often ensconced him. He leaned forward with bulging eyes. "This is going to be a major breakthrough in our understanding of greenhouse gas emissions."

Those words kindled excitement within me. I hadn't known that my work would help to change the scientific community's understanding of methane and global warming. I had been laboring with the faith that my effort would someday be fruitful and important. And now, for the first time, I felt that my trust in my work was confirmed. Maybe this meant that I could become someone important in science after all, and not remain an anonymous nobody the rest of my life.

After our conversation, I drove back to my solitary cabin and smiled as I passed through a blanket of ice fog. I knew I'd been destined for this work, but I wondered, was it simply all my lake love, deriving from happy childhood memories of swimming in Lake Tahoe, coupled with my hard work that was opening a door for me? Or was the hand of the God my mom spoke about, the God I'd been pushing out of my mind for years, also at work? This question was too big for me that day, but regardless, I was eager to delve deeper into the freshwater mysteries that awaited me.

Part II

NOMAD

SIX

Childhood Broken

Mom slammed the station wagon door. Christa and Hanna, my youngest sisters, were wedged shoulder to shoulder, double buckled in the middle back seat. As the bigger sisters, Annie and I sat on either side of them, hugging our knees to our chests. Our feet inside tennis shoes rested on messy piles of blankets, pillows, and hand-me-down clothes. Behind us was a wall of paper bags and open boxes overflowing with kitchenware, hangers, books, and toys—nearly all of our belongings, everything we owned. It was 1986 and another move—I was ten years old, moving into my tenth house, but this time without our dad.

I strained to get one more look at our now-empty Villard Street house, which we'd rented as a complete family for a little over a year from the University of Oregon. Remnants of our firewood were stacked against the side of the house. "What about the wood?" I asked Mom. She closed her own car door and said we'd come back later for the wood. This was good because heating fuel was expensive and having a wood fire in the house really took the chill out of damp Eugene winters.

Our new house, also part of university family housing, was less than two miles away, on Columbia Street. It was two stories, with French windows, and set far back from the curb. From my place in the wagon as we pulled up, it looked like a mansion. Mom said we only had to pay the university eighty dollars a month to live there— a discount—if we agreed to take care of some of their other family-housing facilities while she took classes and waitressed. The first thing Mom said she wanted to do was till the yard and put in a huge garden. My sisters and I wanted to explore the new house. We scrambled out of the back seat, four sets of hands over four sets of feet.

The yard became a blur as we sprinted the long walkway to the door. Mom shouted after us, "Girls, look at the fig trees. There's a cherry tree out back and some plum trees too." We stopped in front of the door, which was located on the side of the house and shaded by a trellis of grapes. Big, round black grapes hung in clusters from the leafy ceiling. We'd tasted grapes before but never had our own at home. Mom had always said they cost too much. "Can we eat them, Mom?" we all shouted, anticipating the taste of the delicious fruit. Four girls jumped and twirled, reaching with fingers outstretched toward the black clusters that dangled above our heads. I pulled one down just as Mom walked up with her arms full of clothes and hangers. "Yes, all of this is ours. You can eat it all!" Being the oldest and with a strong desire for peace and order, I passed a grape out to each sister. We all took a bite. Sweet, delicious juice exploded in my mouth, but there was more. Something grainy, crunchy, and bitter. "Pwaah!" Hanna said as she spat the fruit out partly on the ground and partly on her round four-year-old belly. I looked at the remainder of my grape still pinched between my fingers. The flesh beneath the thick skin was the darkest red I'd ever seen, and inside were five or more large, gray-green seeds. "Those must be the type for making juice." Mom explained to four questioning faces.

Moving was a way of life to us. It was normal. It was all we knew. Yet no matter how many times we experienced it, each new place was still a world of possibilities. Inside the house, the pounding of our feet on hardwood floors echoed from barren walls as my sisters and I dashed around exploring every room, nook, and cranny. Upstairs, Christa spotted a tiny door in the corner of a bedroom. "Maybe it's a tunnel!" Annie exclaimed. Excited, we opened the tiny door. I could see a sliver of light at the far end of a long, dark space. "Yes! It is a secret passageway," I whooped. Despite her protests and because I myself felt a tinge of fear, I made Annie go in first.

Huddled inside the passageway, we heard the voices of strange men and their heavy footsteps on the stairs. We peered through the crack of light as men we recognized from church set down mattresses and assembled bed frames in what would become my and Annie's bedroom. Then we crawled backward, slipping quietly out of the tunnel, and darting downstairs. Mom was just covering the worn red couch with a blue crocheted blanket to hide the blemishes. She put several bright pillows in the corners by the armrests. We ran over and jumped on it.

When the moving men left, Mom unpacked some bowls and plates from the boxes and warmed up chicken vegetable soup. Meat, like juice, was expensive, so we usually ate mostly beans and vegetables and had the free hot lunches served to needy families at school. This night Mom picked remnants of cold chicken meat off the bones she'd saved for herself while we ate the warm soup with slices of her homemade bread packed so full of wheatberries, oats, and walnuts that we were sure it would have sunk if dropped in a lake.

After Mom finished her food, she pulled out her Bible to read to us by the table's lamplight. "By faith Abraham obeyed when he was called to go out to the place which he would receive as an inheritance. And he went out, not knowing where he was going. By faith he dwelt in the land of promise as in a foreign country, dwelling in tents with Isaac and Jacob, the heirs with him of the same promise; for he waited for the city which has foundations, whose builder and maker is God" (Heb. 11:8–10, NKJV).

For us, home had never been a certain place. Places changed too often. In my ten years, we'd lived in army housing in Texas; civilian apartments in Germany; our great-grandma's house in Reno, Nevada; a spare bedroom in another family's home in Reno; a trailer in an orphanage–farm project in rural Oregon; and then the back bedroom in the house of a man who kindly took in our family just outside Eugene when we could afford no other housing. We had started living in University of Oregon family housing when I was six years old, when Dad had entered college on the GI Bill (see center insert photo 15). Instead of a place, home had meant being together as a family. But now our parents were separated, for reasons I didn't understand. Later I was told the reasons were related to differences in faith and money management. All I cared about the night we moved into our Columbia Street house was that Dad was missing from the table. He was somewhere out there in the darkness beyond our new glass windows, parked beside a curb as close to our home as he could be, getting ready to sleep in what would be his new home for at least a year, the camper on the back of his truck.

...

AT AGE TWELVE my bare feet pivoted in a patch of dandelions and I sauntered back toward the open trunk of our station wagon. "Help me carry the food, Katey, or you won't get any dessert," Mom threatened. "I don't want any dessert," I retorted. We were having a picnic at Dorena Lake, and I cared as much about food as I did about obeying. My parents' divorce two years earlier had shattered my world. I had always believed in the righteousness of love. I had grown to the age of ten taking family love and togetherness for granted, thinking of it as normal. Their divorce had opened the door to pain and resentment that chased my former childish joy away. All I wanted now was for them to be together again, for peace and love in our family to be restored. If I could not have that in this moment, on the shores of Dorena Lake, then I would look for a means to numb the hurt. My mind focused on the lake's water, which helped.

Christa pranced and shrieked in the parking lot, beseeching Annie to let her back into the car because her bare feet were burning on the black asphalt. In the trunk, a head of lettuce fell into the potato salad as my mom rummaged through the picnic supplies. Mom shouted frantically at Annie to open the door for Christa. Hanna struggled inside her sundress, a tangle over her head, trying to strip down to her swimsuit. I was already in mine. I hadn't seen the need to wear clothes. They would only be something more between me and the water. I dashed across the hot asphalt and yanked the door open to Christa's and my own burning feet's relief. Then we piled back out of the car, this time in tennis shoes. I dropped my picnic load on the grass, kicked off my shoes again, and dashed into the water as fast as I could. Annie splashed in beside me.

The water was not cold, but standing up to our thighs, it was thick and difficult to move through with our legs. We scooped up armfuls of water near the surface and threw them at each other, screaming and shrieking and laughing as the falling droplets quickly wetted our long hair. Cool mud oozed between my toes. I was fearfully curious about what else might be down there, like crayfish or slimy plants. This water was murky and brown, and it tasted like it smelled and looked. But it was water. The world below its surface was such a great mystery that it alone had the power to make me momentarily forget my sadness, the arguments I incessantly had with my mom, and the shame of my acne-covered face. Swiveling and splashing through the water with my hips and arms, I took intrepid steps across the muddy bottom. Then I drew in a deep breath and put my whole head underwater. My feet pushed off the bottom, and I reached forward with both arms to glide through the cool, silky liquid, my eyes now shut and ears sealed off from all the shouts and sounds of the world above. When my lungs felt as if they were going to explode, I came up for air, then immediately went back down.

"Come eat, Katey," my mom later called out. Just the sound of her voice aroused rebellion inside me. Back on land, when Mom passed me a ham sandwich full of lettuce leaves, I saw the fresh, soggy scabs amidst the lines of scars on my wrist, fingernail scrapes from the last time I had defied her.

At this time, Dad lived with his stepdad, whom we thought of as Grandpa Sid, in a geodesic dome just south of Reno, Nevada (see center insert photo 15). Instead of straight-standing walls with right-angled corners, Grandpa's dome home was a plethora of triangles all fitted together in the shape of a half-sphere shell protruding from the ground. Grandpa had built it himself. In the loft studio, he and my grandmother did art together—that is, before she developed brain cancer at age fifty and moved into a convalescent home. Grandpa's dome was perched partway up a desert slope amidst a grove of Ponderosa pine trees. To get to the front door we got to hike up a path through these trees, climbing timber-framed, granite-sand-filled steps along the way.

The fighting between my mom and me had been so bad the year before, when I was eleven, that I'd left Eugene and gone down to finish sixth grade with Dad and Grandpa in the dome home. As a new girl with large bones, frizzy hair, generic-brand clothing, and overwhelming insecurity, I did not fit in at Pleasant Valley Elementary School. I dreaded recess most, the time of day when other kids clustered together. In Eugene, amidst my own clan of sisters and neighborhood kids, I had always insisted on being the leader. I had decided what we would play and how. I hadn't always found the play to be much fun when I was in control, but I didn't know how to play without being in control. A structured environment was a different story. I had always thrived within the social bounds of my soccer team or gifted-and-talented math class, but it was the unstructured situations that I felt I needed to dominate or die. There was no in-between. In Nevada, in a group as large as my new school's, I was far from in control. I would stand somewhere on the concrete playground hoping no one would notice that I was alone. My depression, rooted in the pain of my broken family and in my own self-consciousness, was exceedingly deep. To evade school, I feigned sickness, but the long days alone in the silent dome home were only worse. There was no adult to tell me I couldn't watch television, so I did, but when I flicked it off and stared at the triangular panels of the walls, depression was still there.

One night, in the dark after dinner, I slipped out the back door and raced down the wooden deck stairs. I ran into the pitch-black and up the hill behind our house, one of the thousands of desert foothills of the Sierra Nevada. Dad opened the screen door and called after me in concern, but I kept running. My legs pushed through sagebrush and thistles. My feet stumbled on rocks, but I didn't fall. My heart pounded. I ran and ran. Dad chased me up the hill, calling all the way for me. Somewhere, a thousand feet or more up the hillside, I tripped and fell. Standing back up, I raised both arms and my face to the sky, and cried to God from the bottom of my heart to please take my life. My innocent trust in what seemed

Alpine lake

good and right, a unified family, had been shattered. As a child, I had believed in my parents' love for each other when they kissed in the kitchen; my sisters and I had giggled and turned our eyes away, smiling in embarrassment. Their divorce had driven a thorn into my heart that was too much to bear. Was there any way this pain could be undone? I believed up on that hill that God, for whom nothing is impossible, could take me up to heaven, then and there. But he did not. Dad caught up with me soon enough and held me in his arms while I shook and sobbed.

...

THE ONLY SOLACE I found at that time was in the Sierra Nevada. When school let out six months later, I sat in the middle of the bench seat between my dad and Grandpa Sid in Grandpa's small tan pickup truck. His blue canvas backpack, which contained our lunches and water canteens, jumped around on the floor as we jolted our way up the winding dirt road higher into the mountains. The windows were wide-open. Fragrance of pulverized pine cones and pitch wafted in, and I could hear the splash beneath the tires whenever we crossed outwashes and small creeks. We pulled to a stop on a level area of granite stones where the road ended and a weathered rough-cut pine fence separated us from a lush, green pasture and an old, brown log cabin. Cows grazed on either side of a bubbling, clear creek. Lacing up our leather boots and donning our sun hats, we set out up the first mountain valley pass toward Schneider Cow Camp. Since settlers first arrived in the Sierra Nevada region, it has been the custom to let cattle graze in the mountain pastures in summer. Sometimes if we stood very still and if the wind was not blowing too strongly through the pines, we could hear cowbells clang through the valleys, echoing off the walls of granite.

Dad's and Grandpa's long strides quickly separated them from me along the trail. As a youngster, I'd learned that crying out for them to wait was useless. Grandpa always said that everyone walks at his own pace. When the time was right, they would wait.

It was so peaceful up in the mountains. Nobody was up there but us. The air was dry, and there wasn't a cloud in the blue sky, which seemed so much closer at this elevation. Red powder from the dusty trail coated my shins. Around me, on either side of the trail, June wildflowers bloomed—red Indian paintbrush and purple lupine. I knelt and touched the fuzzy leaves of lamb's ears and savored the smell of sagebrush baking in the warm sunshine.

Near the shoulder slope of the hill, I caught up with Dad and Grandpa, sitting under the shade of a pine.

"Katey, how do you think that rock got split so perfectly in two?" my dad asked.

I looked up at the twenty-foot-tall granite boulder in front of us. It looked to me like a giant had walloped it with a rock machete. "I don't know," I said.

I turned to my dad. His hands clasped a canteen between his folded-up lanky legs. He had that dorky, open-mouthed expression that seemed to come from restraining a big grin of eager curiosity. My dad was a computer programmer for a casino gaming company called Bally Systems. He'd been raised by his stepdad and his mother, never knowing very well his own father, a pathologist with the Armed Forces Institute of Pathology and the National Institutes of Health in Washington, DC, and Bethesda, Maryland, respectively. While growing up, he had felt he was an inconvenience to his stepfather, a successful Reno realtor. His mother had been a beautiful artist who loved to sit in the mountains and paint. She had been so engrossed in her painting one day that she didn't notice my toddler dad fall into a creek and nearly drown. A stranger came to the rescue, and my dad's mom seemed startled to realize that he was even there. After my parents met in college, they became part of the 1970s Jesus movement. My dad's atheist mom and stepdad strongly disapproved of this and even more of my parents' marriage.

Soon after, Dad joined the army as a medic, with the dream of one day becoming a doctor. But the struggle of feeding a burgeoning young family led him instead to get a degree in computer science at the University of Oregon. It was there that Dad read books and spent time critically thinking to a point at which he denounced his Christian faith. When I was seven years old, he began planting seeds of doubt in all our minds, purporting that Christianity was just another fairy tale.

Here on the hiking trail, he was now planting another kind of seed in my mind, a seed of curiosity about science. Dad's head, covered in wavy dark tassels of hair, was cocked to one side. His hazel eyes gleamed behind his glasses as he anticipated his explanation of the dissected boulder towering above us. Then he told me about the power of water. "Rain makes its way into tiny cracks and pore spaces in the rock. Later, in winter, when the air temperature falls below freezing, the rainwater turns to ice and expands. The force of the water expanding as it freezes is so strong that season by season it has the power to split rocks, even rocks as large as garages." I climbed up on that rock and looked down through the crack at my dad and Grandpa below.

We drank cool water from our canteens, then ascended the saddle overlooking Schneider Cow Camp. In the valley below us, the sunlight reflected off a series of sapphire-blue lakes, and we could faintly see in the far distance the southern end of Lake Tahoe. We traversed a field of melting snow, its edges laced with the brown, decaying leaves from last summer's vegetation. Bright yellow-green shoots of this spring's new growth skirted the dead leaf matter. With each quarter of an inch distance from the receding snow patch, the vegetation became a deeper green and progressively taller until a full carpet of wild flowering plants extended as far as I could see across the mountain's flank.

We jumped over the babbling creek flowing with meltwater. The creek emptied into a lake, where we stopped to eat our lunch. Before lunch, my dad disappeared behind some rocks, and the next thing I heard was his whoop and holler as he dove into the crystal-clear, frigid water. That sounded like an ecstasy I didn't want to miss out on. I stripped down to my underwear and waded into the lake. It was as cold as ice. Goose bumps jumped out on my skin. "If Dad can do it, I can do it," I whispered to myself. But when I dove beneath the surface, I thought my lungs had frozen shut and I wasn't sure I'd be able to breathe again. Resurfacing, my tight lungs gasped for air. I turned around and reached for the shore. As I sat dripping on a rock, the sun's energy slowly dissolved the burning cold feeling from my skin. The lake's water was clear, and I could see immense boulders many feet beneath the surface. Sunshine glinted off ripples in the middle of the lake. It was good to be out here, where the serenity of nature seemed to separate me from my cares.

***

IN JULY OF that summer, I promised to get along better with my mom, so my parents let me move back to live with her and my sisters in Eugene. But the incessant sounds of shouts and slamming doors permeated our home. In August, I started seventh grade at Roosevelt Middle School. I rode my bike across Eugene on Saturdays to clean the church. This and the monthly seven-dollar home permanents I learned to give to several ladies we knew were the start of my own savings account. I kept my money in a pink porcelain pig, and by mid school year, I'd earned nearly sixty dollars! I was afraid to spend a cent; it was too valuable. This was real green, paper money, not the food-stamp coupons I was so humiliated by when Mom pulled them out of her wallet at the grocery store. Food stamps were almost as embarrassing as the prayer requests Mom announced in church for our single-parent family's need to make ends meet.

"Save your money, Katey," Mom would tell me. "If you work hard in school and get straight A's, then someday you can get scholarships to go to college. Don't make the mistake your dad and I did, getting married young and having children before going to college."

One day we pulled into our driveway and instead of opening her door, Mom put her face on the steering wheel and started crying.

Her shoulders heaved with her sobs and drips fell to the floor from her face. My sisters and I sat like statues in our seats. I didn't know where to look, but I could not bear to watch her.

Mom lifted her head. "I just don't know what we are going to do," she said. "We don't have any money."

"I have money, Mom," I said. "We can use that."

"I don't want to spend your money, dear Katey," Mom answered tenderly. "That is yours."

As we walked up to the front door, we noticed a white envelope stuck into the screen. Mom took it down and opened it. She reached inside and gasped. "A hundred dollars!"

Our eyes got big. "Where did this come from?" we all asked at once.

We never did find out, but our mom's big burden was relieved. We also took this as a lesson that there are kind people in the world, maybe even angels. I promised myself that when I grew up, I would do everything in my power not to be poor.

...

HALFWAY THROUGH THE seventh grade, I left home to live with a family from our church. A strong-willed twelve-year-old was too much for Mom on top of her college classes and nighttime shifts as a waitress. Besides, the arguing was not fair to my three younger sisters, who cowered as Mom screamed when my cool, calm words defied her. I loved my family, but I wanted to get away. I needed to leave. I had a blind sense that I could go further in the world outside the walls of chaos and dissention.

The family from church was quiet. Shockingly few words were spoken in their home, and I never heard a single raised voice. The Howes had moved to Eugene from South Dakota and often spoke with nostalgia about their former life on the prairie. The handsome, blond father shaved his legs to race bikes, and I thought the mother, with her strong jawbone, dark eyebrows, and long black hair, was exquisite. She was part Native American. I had always dreamed of being a Native American princess, living off the land with my people. The Howes' home was as close as I ever got to living with Native Americans, but our life was far from wild. I shared a bedroom with the oldest of their three girls and was expected to be at the breakfast table by 7 a.m. sharp each morning. This family could afford boxed cold cereals and even allowed some sugary options, like Cocoa Puffs. I tried Cocoa Puffs once, but preferred the unsweetened options, which were more similar to the boiled wheatberries I'd been raised on. Knowing that my dad's monthly two-hundred-dollar child support payment for my care now went to this family eased my sense of guilt for eating their food. I didn't dare offer to spend any of my own, precious money, now a savings of nearly a hundred dollars, for fear it would disappear faster than I could renew it.

One morning at the Howes' house I was startled awake when I heard my name whispered. "Katey." I opened my eyes and turned my head in the direction of someone gently shaking my shoulder. It was my mom. She was kneeling next to my bed. "Katey, your sisters and I are moving down to Reno. Dad and I have decided to get remarried. Would you like to come with us, or do you want to stay here for two more months to finish out the school year?"

"Remarried?" I asked in a groggy whisper. My eyes met hers for a moment and then I closed them. "It will never last. I'll stay here." With a new thorn in my heart, I turned to the wall. Mom got up to leave. "I love you," she whispered, and the sadness in her voice bounced off my heart.

...

THE SUBURBAN RENO home where my parents set up shop had a small patch of grass in the front yard. The landscape consisted mostly of asphalt and concrete poured on top of the desert to accommodate a sprawling population. Without a tree in the neighborhood, the place felt lifeless.

When I joined my family after school in Eugene let out for summer, I found that my parents still argued. They argued about money and they argued about God. "Girls, Christianity is a fairy tale," Dad challenged us. "Why would you want to believe in a fairy tale?" Mom would cry, "Tom, please don't say that to the girls." But it was too late. The seeds of doubt had long ago been planted.

Despite this spiritual storm, my dad's presence in the house restored the balance of emotional control and rationality that I had tried and failed to provide in his absence. With him there, I had less reason to fight with Mom. On the contrary, she and I took long walks at night through the desert streets. She told me about how she almost became an exchange student in Norway, and about her affinity with all things Russian—the people; the tragic stories of oppression, and the strengthened church in the midst of it; Dostoyevsky, Tolstoy, Tchaikovsky; and Russian ballet. She told me about communism and how the state regulated the economy so that everyone lived in exactly the same kind of house and had the same amount of money to spend. This brought to mind the village of milk-carton houses my third-grade class had constructed in Eugene. Each house had been painted differently, but otherwise, they all had been the same. How I had longed for the differences between rich and poor to be leveled because I was ashamed to invite rich kids home from school to see our house. Having never beheld pictures of the Soviet Union, I wondered if the people did not live in some kind of giant milk-carton-village world where kids didn't suffer from disparities.

One evening during my freshman year of high school, I showed my mom a flyer I'd picked up at school. There was going to be an information session about the American Field Services student-exchange program, and I wanted to know if she'd take me. She would!

The following week we arrived at a tiny auditorium at the University of Nevada's Reno campus. We took two seats near the back, where, from the top of the sloped classroom, we could get a good view of the slideshow being presented. The lights went out, and the screen lit up with an image of dark-skinned kids playing soccer in a tiny village perched high up in the Andes Mountains. To get there, the speaker had traveled on a crowded bus first through thick rain forest and then into the steep alpine cloud forest. I gasped when the carousel advanced forward to show the bus broken down and perched on the narrow ledge of a winding road. People stood around the bus, and beneath them, red soils and rocks gave way along the precipice, dropping more than a thousand feet.

By these slides, the world suddenly became bigger and more accessible than I'd ever known it could be. The speaker had been there. For the first time, I felt hope lay out there somewhere far beyond the paved suburbs where I had become trapped. A desire ignited inside me. A dream was born to get out of my life, to go somewhere beyond the bounds of my unthriving family. I burned for an entirely new world, fresh with opportunity.

The American Field Services was accepting applications to spend a year abroad studying. I could participate, if only I could come up with the five-thousand-dollar program fee. At the start of my sophomore year in high school, my family moved to a suburb of Las Vegas without me. I petitioned to stay behind in Reno. I was doing well in high school and sports, had made good friends and didn't want another move to disrupt my academics and future. Money was my only concern. It always had been. Since Dad always held on to his tightly and Mom never seemed to have any, I knew it was up to me to find a way to survive. I rode my bike everywhere and got jobs as a bagger at the grocery store and as a janitor at a hair salon. Meanwhile, I worked hard on the application paperwork for the American Field Services student-exchange program. Several months later, I was thrilled to discover that I'd received a scholarship to go to the former Soviet Union.

SEVEN

A New Beginning

The year was 1992, and I was a sixteen-year-old exchange student in Russia. I had been in Krasnodar, a large agricultural city in the breadbasket of southern Russia, for two weeks. So far I had ventured only beyond the gate of my new home with my host family. They were nice enough: the mother, Ira, and her teenage daughter, Kristina, were the two members of the three-generation Armenian immigrant family I spent the most time with. All I had done in Russia was go on family errands around the city and wait for my Russian school to start in a few weeks. The truth was, I was struggling with the culture shock of Russia: their home and the city were frightfully different from anything I'd known in the US. My host family didn't know any English, and my ignorance of the Russian language and fear of getting lost didn't help my adjustment to this new place. Two weeks into this yearlong adventure I'd signed up for in an attempt to get away from my family and nonexistent home life, I needed to get some time outside. Alone.

Accustomed to strenuous daily exercise on sports teams and bicycle commuting in the US, I needed to go for a run. I put on my running clothes and ran in place in the kitchen, trying to show Ira what I wanted. Since jogging was a foreign concept in this culture, I doubted that my charades helped. Luckily the word "park" is the same in Russian and English. Ira thumbed through my dictionary to show me the word *ozera*—"lake." A lake! I pictured the pristine blue waters of Sierra Nevada lakes as Ira waved in the direction of the park with a lake. "*Da* (Yes), a lake!" I said, grateful, and excitedly smiling at her. She smiled awkwardly back, using her lips to try to hide her mouth dotted with golden or missing teeth.

My heart pounded as I stepped outside the gate. I was familiar with the first half-block, to the communal water pump where Kristina and I fetched buckets of water for the family's daily use. The neighborhood was less than a mile west of Krasnodar's main avenue, which was lined with government buildings, theaters, and high-rise apartment buildings, where Ira and I had walked a handful of times before. But now, at the water pump, I must turn east, a direction we had never gone. To keep from getting lost, I followed the tracks of the tram, which curved past a landscape of densely packed single-story houses, until I reached a lake skirted by green grassy banks and some shade trees. People sat on benches, resting and enjoying a picnic. Local babushkas, their heads covered in colorful kerchiefs, leaned toward each other, gossiping. Young families with children bathed in the sun and splashed in the water. Under the shade of a large willow tree, two men fished. They didn't seem to be bothered by the pea-green color of the water or the plastic garbage and other flotsam on the surface.

All of these people stopped what they were doing and stared at me. I knew that my white T-shirt, shorts, and clean, cushioned running shoes looked very different from their dusty, drab, and well-worn attire. Likely none of them had ever seen an American before, let alone an American running through their park. I knew

I was a spectacle, but I didn't really care. It felt so good to run. I ran hard. My knotty hair, gathered in a tangled ponytail, swished across my back with each pounding step. Sweat streamed down my temples and stung my eyes. I lifted the bottom of my T-shirt to wipe the sweat away but stopped suddenly in my tracks on the far side of the lake. There, floating in the water, was a dead dog. It was a red-haired, medium-size dog now hugely bloated and covered with flies. My stomach turned somersaults as my eyes passed to the dog's head. Its eyes were bulging out and its mouth was open, black lips exposing its sharp teeth. The Russian people did not seem to take notice of the dog, but they were watching me, and I tried to hold back my urge to vomit.

I did not make another loop around the lake. Instead, I found the road with the tram tracks and ran straight back to the corner where our water pump stood. There I stopped to catch my breath. On familiar ground again, I did not bother to run or look around. I wanted only the respite of getting home. Yet, as I walked, in my homesick heart I was painfully aware that the single-story house engulfed in a concrete courtyard half a block away was not home. I sauntered toward the gate. Suddenly an arc of lukewarm liquid splashed across my shins and shoes. I turned my head to see a man wedged between the broken seams of the neighbor's welded fences, smiling sheepishly at me as he started to close up his pants. Not knowing whether to cry or throw up, I slipped back into my host family's courtyard and didn't venture out again on my own for another two weeks.

...

THE START OF my Russian school brought a new set of challenges. Language was my main barrier, not only to coursework but also to navigating an unfamiliar social scene. Two weeks after school started, I accepted an invitation to go to the movie theater with some classmates. Unfortunately, I hadn't yet learned the word *uzhasni* when I accepted my classmates' invitation. All too soon, I realized it meant "horror." For two hours I sat looking down at the chair in front of me, hoping my classmates wouldn't notice my strange postures as I diverted my eyes from horrific characters, blood, and fear on the screen in front of us.

After the movie ended, we walked down the wide central promenade that stretched from the theater toward the city center. Soviet-era flower boxes and bushes lined the sidewalks, and groups of people turned off along obliquely oriented paths radiating out in different directions. Suddenly a dark figure holding an AK-47 emerged from a bush we were passing. His voice was cool as he ordered us to lift our hands. Without thinking or waiting to see what would happen, I dashed off behind the bush. My heart pounded and I wondered if I'd be shot in the back as I ran. Without ever looking back, I made it home, panting and out of breath. Ira scolded me for arriving home alone in the dark, without an escort. "This is not Soviet times anymore, you know. Russia is dangerous now."

She sat with me at the kitchen table as I ate the fried egg and noodles she'd had waiting on the stove. I stared at the pot of concentrated black tea and the sugar bowl. Then, thanking her, I got up and walked through the main room hung with thick dark Armenian carpets, past the babushka snoring on her bed and my host father smoking in his underwear in front of the television. Kristina's bed was empty. I gathered that she must have snuck out again and was probably smoking and drinking somewhere downtown with her friends. I crawled into my own bed, snuggled deep under the duvet cover filled with a Soviet wool blanket, and laid my head on a large square pillow and cried. I'd lived in other people's homes before, but all of this was just too much. I wanted to go home, back to the US, even back to my parents. But I had no way to communicate with them. The family had no telephone, and letters took months to transit, if they made it there at all. I pulled out the gray, leather-bound copy of the Bible my dad's real dad had gifted to me before I'd left for Russia. At least the Bible was a familiar book. I opened to Psalm 32:7 and recognized the words from a song I'd sung in church many times growing up and in the back seat of the car with my sisters while we'd waited in the parking lot for our mom to grocery shop. "You are my hiding place. You always fill my heart with songs of deliverance. Whenever I am afraid, I will trust in you" (adapted). Trusting in God was all I could do, and my sliver of faith brought comfort.

...

THREE MONTHS INTO the school year, I was still struggling but had made one notable improvement. All the tremendous mental energy I was accustomed to giving to my coursework in the US was put instead into learning the Russian language. I wrote vocabulary words on small pieces of paper and tiptoed around the house memorizing them while others were sleeping so that I could put the words to use each new day. But despite my efforts, my language skills were still far too limited to have any hope of keeping up with my peers in precalculus, chemistry, history, and literature. Further improvement in my new life came when I began working with a private teacher. Irina Mihailovna, a professor at Kuban State University, was a short, plump, middle-aged Russian woman with curly gray-brown hair and a large bosom tucked inside a starched, white-lace, fully buttoned blouse. For approximately ten US dollars, which I exchanged illegally and nervously each week into rubles on the black market because I could get a better rate than at the bank, she gave me six hours of lessons a week in Russian as a foreign language.

Still, with every step forward, I seemed to take two giant steps back. One fall afternoon, Irina Mihailovna met me at the tram stop outside the university and escorted me onto campus in a tattered pair of high-heeled shoes. I traipsed behind her in my tennis shoes, past stern-looking guards at building checkpoints and throngs of college students. To my surprise, I found that learning the Russian language was turning my world from black and white into color. I began to discover that Russian was logical and rich. With one word, Russians could express a thought requiring a string of words in English. For instance, *temneet* means "it is growing dark outside." This word also seemed to express peoples' feelings for the state of life in their country, where the economy was highly unstable and food hard to come by after the fall of the Soviet Union.

Irina Mihailovna had taken me once to the art museum, an elaborate pink building on the corner of the city's main avenue, embellished on the outside with white sculpted moldings and crowned turrets. Inside, my teacher helped me read the titles of pre-Soviet Russian paintings of peasants and asked me to describe to her what I was seeing in them. I knew that Irina Mihailovna had translated Shakespeare in her past, so I inserted English in places where my Russian fell short, hoping she'd teach me the Russian through translation. But she refused. My English was met with blank stares, and then she patiently pulled from me what Russian I knew and gently added more to it.

When we entered the university's auditorium that day, Irina Mihailovna began to show me that normal, simple nouns, such as the word "house," changed in Russian depending on how they were used in a sentence. I'd learned about verb conjugation in English grammar, but I could not fathom that nouns could change form too. There must have been something wrong with what I thought she was trying to tell me. Seeing I was confused, she started declining a different noun, the word "car." Then "man." The rigidity of my mind exploded. In utter frustration, I stood up, gathered the open textbooks, notebooks, and pencils from the table, and sent them flying to the floor. I raised my hands over my head and shouted, "*Ya ne ponymau* (I don't understand)! Why won't you just explain this to me in English?" My cries ricocheted off the walls of the immense auditorium. I had surprised even myself. It was not like me to lose my temper. I felt so out of control in this foreign culture, and I was starting to not recognize who I was anymore. At just sixteen, I had tried to essentially run away from home by going to Russia. But I wasn't feeling any more at home with myself here, and now I was also ashamed at my outburst. Irina Mihailovna sat still and silent, her expression unmoved.

...

I HAD REACHED my lowest point in Russia. Trams frequently broke down. I was used to piling off one broken tram with a stifling crowd of other passengers to wait for a replacement. But one day on my way to meet Irina Mihailovna, my tram stopped just short of the university. There was no announcement that passengers should get off. Just a very long pause in motion. Eventually passengers started shouting and pounding on the doors and windows. Finally, registering the cries, the driver opened the doors, and I squeezed out among a torrent of shoving people onto the grass. Then I saw it. In front of the tram on one side of the tracks lay a man's corpse. The head had been severed and lay by itself between the tracks in front of the tram. Men drinking kvass, a fermented drink made from stale bread, left their communal cup at the large barrel on the roadside and joined the crowd gathering around the body. I did not join the commotion, or stay to find out exactly what had happened, but hurried along to the university, eager to find safety and consolation in Irina Mihailovna.

When I told her what I'd just seen, Irina Mihailovna shook her head, and the sadness that was almost always present in her eyes grew more intense. "*Uzhas* (Horror)*! Bozhe moi* (My god)*!*" she exclaimed. I didn't think for a minute that she or any of the other hundreds of women who called on their god actually believed in any god. For too many decades the Soviet Union had been their god, and like all idols, it had finally crumbled. The people of Russia were left empty. Their souls were as barren as the shelves in the stores. Stores were places where people stood for hours in long lines with food ration tickets, only to find themselves arguing with a saleswoman behind the counter when they reached the front and found that the last of the bread, rice, or milk had been sold. They were told to move on, to look in another store in another part of town. And so the hungry people moved on, their empty cloth sacks dangling from stooped shoulders and withered hands.

...

"*PRIVIET* (HELLO)," a tall boy said to me one day. He moved closer to share the handrail where I was standing on the bus. "My name is Valera. I saw you running last Sunday, down by the river, where the people were gathered."

My mind shot back to Sunday. In spite of my awful first run in Russia around the lake, I had picked it up again because I had too much energy inside me not to run. As my language skills improved, I gained confidence in exploring other green spaces in the city. That particular Sunday I had been out for a jog just after sunrise but stopped when I saw a crowd of people gathered along a grassy bank of the Kuban River. A man had been standing chest deep in the water, holding his hand over the head of a woman. She trembled in the water next to him. "I baptize you in the name of the Father, the Son, and the Holy Spirit," the man had declared, lowering his hand to her head and dunking her backward in the water. The woman emerged from the river, dripping with water and crying for joy. People on the shore, who I thought must be a church congregation, enveloped her with dry clothes and warm embraces, uttering praises and hallelujahs. Then a different man waded out into the water. I watched as tens of people in the crowd were baptized in turn. On shore, others were singing hymns. I felt like I had stumbled into a story from our family's children's Bible, the page with John the Baptist.

Until that morning I had given up hope of finding a church to go to. Despite Dad's challenges to my faith, I had not stopped believing in God and had gone to church often on Sundays with my mom in the US. I wanted to go to church in Krasnodar because I hoped it would bring some element of familiarity to my unbearably strange new life in Russia. When I'd asked Ira about church months earlier, she walked me downtown to a place where the city's old Russian Orthodox Church was being reconstructed now that religion was legal again. Black-robed priests with long dark beards wove their way between scaffolding that covered the building inside and out. New onion domes were being hammered into place at the top. At the entry, a babushka sold candles for people to place in front of icons. This was all very culturally interesting, but orthodoxy was not what I had in mind. I yearned for something more familiar, people who expressed their faith the way I was accustomed to at home. These Baptists at the river had offered a new glimmer of hope.

"I was with the people at the river," Valera said. "I saw you running. I like to run too!"

A grin spread from ear to ear across his thin, chiseled face. His large blue eyes smiled too. Valera's hair was cut into a Mohawk, and waves of thick blond hair fell down beside his cheeks. He tossed his head back, gathered the hair into a ponytail, and bound it with the bracelet he'd worn on his wrist.

"I'm Katya," I said. "I usually run through the parks along the river, but I've never seen anyone else running." I thought of all the Russians who had laughed at me and said that the only time they run is to catch a bus. "Where do you run?" I asked the boy.

"Out near where I live in the garden region," Valera answered, still smiling. "You are a foreigner, an American, right? How is it that you speak Russian? I'm glad you do because I don't know any English."

For the first time Valera's face showed a hue of remorse.

"I haven't run in a while, and I'd really like to run again. Should we go together?"

"Sure!" I said. I couldn't believe my ears. I'd never gone running with a boy before.

We agreed to meet at 5:45 a.m. the following morning at the stop where the bus line intersected the tram tracks. My heart beat madly with so much excitement that I could hardly sleep that night. I didn't want to miss my alarm or be late.

I usually ran in the mornings but always after sunrise. This day would be different. It was dark when I left the house. Still, I knew where the curbs and broken sidewalk tiles would be since I had navigated my way in the daylight many other mornings, weaving my way between people leaving for work and the babushkas setting up cardboard boxes along sidewalks where they would sit all day selling keys, bundles of cilantro, or other small items. I found Valera at the appointed meeting place. He was stretching his leg on a metal beam.

"Good morning," he said, smiling again and swinging his leg down to the ground. "Shall we go?"

"Aren't you concerned about running on concrete in those shoes?" I asked him, staring incredulously at the generic Keds-style shoes he was wearing and his long shorts, despite the rather chilly February morning. "These are my new shoes," Valera said, twirling his feet around with an air of pride.

Valera set the pace and direction. I struggled to keep up with him. He ran along the tramline and then turned through a rusty metal gate I'd never before entered. We found ourselves in a park complex surrounded on three sides by an oxbow of the river. We ran down a long, straight sidewalk, leaping over tall grasses that had pushed up through cracks in the concrete. I saw a series of rusty, abandoned food carts and overgrown gardens with dry fountains. Surely at one time, this must have been a beautiful park where people could come to stroll and relax. Huge swan boats sat on the shore shaded by trees and partially covered with dirt and vines. Now nobody was here except us. Walking paths radiated out in many directions. In one area, we saw tennis courts. The nets still hung in place, though they were tattered and frayed. Valera didn't seem to notice the disarray. "I could find some rackets and we could play tennis," he beamed.

I loved tennis. "Sure," I said. "That would be great fun."

We stopped at a playground somewhere deep inside the park. What must have once been brightly colored equipment was now a series of rusty bars where the yellow, red, and blue paint had chipped away. Valera swung up onto one of the bars and then did a somersault flip to the ground. Next, he went over to a vertical pole, put his head against the pole, and gripped the pole on either side of his head. With his elbows bent, he kicked his legs and body straight out. There he was, a straight line hanging out perfectly perpendicular to the pole in midair! He held this pose for a moment, then jumped down and grinned. I didn't mind his showing off. I'd never seen such stunts before, and besides, this gave me a minute to catch my breath.

"Wow, it doesn't seem to me that you're out of shape," I said.

"It feels good to move again," Valera answered. "I'm a dancer. I've been recovering from a blood transplant. One evening, about two months ago, I set off for home after performing a street concert downtown. I was walking alone when I got knifed. By the time someone found me, I had lost nearly 40 percent of my blood and my shoes were stolen. I had to get new blood."

"And new shoes," I said, nodding to his off-brand Keds.

Valera looked at his watch. "I can run you home, and then I need to wash up and get to work." Valera said that he worked as a cook in a pizza restaurant in the city center. Instead of finishing high school, he'd opted for a year at a cooking college. "Maybe sometime you'd like to come to my restaurant and I could make you a pizza?" Valera asked with the bright gleam I was starting to recognize as characteristic of him.

In those days, without telephones or computers, a face-to-face agreement was how plans were made. We would meet every day at 5:45 a.m. As spring turned to summer, we were not alone at the river. A small club of grandmothers and grandfathers began to meet on the grassy bank to stretch and swim in the early morning sunrise. "Now, that looks like the right way to grow old," I said aloud in Russian.

In the months that followed, we added to our morning outings long games of tennis and visits to the gymnasium where Valera had friends training for the Olympics. They showed us their stunts and let us jump on their huge rectangular trampolines. One day Valera took my hand as we climbed off a tram. "Katusha," he asked, using the diminutive form of my name, "do you think we could pray together when we run in the park in the mornings?"

"*Da, konyeshna* (Yes, of course)*!*" I had a hard time containing my excitement. Finally I had met someone in this country with whom I could share things I cared about, things I was curious about, things that made me feel like an authentic version of myself.

The next morning, toward the end of our run, Valera took my hands in his again. We were standing on the crumbling concrete steps by the exit from the river park. I held his hands and closed my eyes. I'd never prayed in such an intimate way with anyone before, especially not a boy. Spiritual life surged in us. We prayed in turn, him in Russian and I in English, uttering thanks to God for all that he had given us.

Valera introduced me to Andrei and Kostya, his two best friends from church. All three of them had become Christians within that year. I went with them to Sunday morning church service, held by the same group of people I'd seen at that first morning baptism at the river.

These Christians were different from all other Russians I'd met so far that year. I could pick one of them out on the street because instead of wearing sorrow and a scowl, their faces shone brightly. It didn't matter how tattered their clothes were or how many teeth were missing in their mouths. Light emanated from these people. They had an ever-present joy that I had not seen in any other Russians, and not even in other Christian Americans, come to think of it. Their light was a testament of the light of Christ I'd been reading about in my Bible. I wondered if maybe Jesus Christ was the hope for *svetleet* ("it is growing light outside") in this dark post-Soviet land.

When I first met these Christians, they gathered for services in a little wooden building, several miles' walk beyond the last bus stop on the edge of town. The pastor of the church had been an underground preacher during Soviet times. Unwilling to denounce his faith, he'd been arrested by Soviet officials and sent to prison, separated from his family for twenty years. During the Soviet era, his family and a handful of other believers continued to meet secretly in underground house churches. Now they were free to worship in the open. Neighbors heard the singing and joined, first out of curiosity and then in newfound faith. In the months that I was there, the little wooden neighborhood church swelled with people. In order to accommodate everyone who wanted to come, they were forced to find a bigger building. Sunday morning services were soon held in the main drama theater in downtown Krasnodar, still an hour commute for me, Valera, Kostya, and Andrei but in the opposite direction. We smiled at one another as we clung to the handrails inside the tram, trying to keep our balance as the train clattered down the tracks. My heart was happy and full. I felt more at home in these moments than at any time I could remember in all my life.

During the week, my friends and I went together to a home Bible study. The gracious hostess always served cakes and tea and grapes picked from her garden. Her husband, an artist and unbeliever, remained outside in his garden while we had Bible study upstairs in his house. The evenings grew long as we visited and sang with the accompaniment of Kostya's guitar. Long past dark, Andrei, Kostya, and Valera would escort me home. One night we stopped on a square near my host family's house. Kostya asked if I could dance like a Cossack. He shot down into a squatting position and then started singing and kicking his way around the square, lifting one leg at a time up above his head to propel himself forward. Valera quickly joined him, adding extra jumps and flares and shouts. Andrei stood stiffly watching and laughing while I attempted one or two kicks before my legs failed me. My host family's house was well out of their way, but escorting a girl safely home was important to Russians, and for me, I felt loved.

...

WHEN I FIRST arrived in Russia, I had fantasized about going to the legendary Lake Baikal. I had experienced so much newness in those first months: a new language, a new community of friends, even a new faith, but my love of lakes was old and familiar. I still had hopes of making a trip to Lake Baikal before the year was up and the time came for me to leave Russia. The director of the exchange program had promised the possibility of it, but with only a month remaining, it became clear that this would not take place. Seeing my disappointment, Valera suggested that we go backpacking at the Black Sea instead. We packed our rucksacks and canvas for a lean-to, and bought tickets out of town. Our tickets must have been especially cheap, maybe the ones normally sold to the itinerant Romani people, because instead of sitting on seats, we were told to ride on the large piles of bundles and bags that bounced up and down in the back of the bus.

I tried widening the slit in the dusty cloth curtain to peek at the world outside. Mostly what I saw were agricultural fields and highways, but after several hours, I caught a glimpse of a steeply sloped vineyard with a little house and garden. Soon the sea itself could be seen, the Black Sea that was actually quite blue! Our bus twisted and turned for at least another hour before Valera and I got off and began our hike up into the hills bordering Russia and the Republic of Georgia. For several days we climbed through the hills. I knew my time here was coming to a close, and I wondered when I would ever see Valera, Kostya, or Andrei again. "God knows," Valera said, and I knew he was right.

The late-spring days in the mountains surrounding the Black

Sea were already long and hot. As we descended from our hike, the shimmering blue water came within view. I ran out of the hilly scrubland and across the sand. Kicking off my shoes and slinging my rucksack to the ground, I sprinted down a wooden dock to leap into the water. It was only when I reached midair that I noticed the water was teaming with huge white jellyfish. It was too late to turn around. I shrieked, hit the water, and somehow managed to take what seemed like the same trajectory backward up onto the dock, unscathed by the thousands of long white tentacles that pulsed in the water below me. Valera's mouth gaped in a tremendous grin. Loud laughter revealed his widely spaced teeth. "*Katusha!*" he shrieked, tousling my wet dripping head. "Don't you know, you have to look before you jump?"

On my last day in Krasnodar, Kostya, Andrei, and Valera fulfilled their promise to take me on a *pivnyak* (beer drinking outing). With Kostya's guitar, a sack full of marinated pork, matches, and a large bottle of beer, they led me out of town. We walked past the last of the high-rise apartment buildings, where Andrei lived, through long stretches of faceless storage garages to a place where the Kuban River sharply bends. Downstream across the river we could see the rooftops of a village. Those were the houses of the Adyghe, Kostya explained. They are an indigenous people, not friendly to Russians.

While Andrei and Kostya made a fire and began roasting pork, Valera and I dared each other to swim across the river. We calculated that we could make it across before the current carried us to the Atagai. We did make it, but we arrived closer to their village than I would have liked. Climbing out on the far shore, we ran as fast as our bare feet would carry us back up along the muddy riverbank. We ran past the point where Kostya and Andrei were roasting meat on the far bank, but not far enough to see what was coming toward us from around the bend. Halfway through our return swim a horn sounded. Long, loud blows followed by short blows. The noise came from a huge barge headed straight for us. Valera shot ahead of me. The horn continued to sound as the barge drifted closer and closer to where I was left alone in the water. I looked up hopelessly at my friends, who were frantically jumping and shouting on the shore. I swam as hard as I could, wishing that somehow I could move more efficiently through the thick water. Just in time I reached the shallows near the other side as the wake of the huge boat rolled over me. "That was a close call," Valera said, giving me a hug (see center insert photo 16). "You almost got your wish to stay in Russia."

...

MY BAGS WERE already packed for the return trip to the US. Most of the clothes I'd brought over with me I left with my host family. The day before I was slated to depart, I told the family goodbye and walked away with Valera carrying my US Army bag over his shoulder. Our relationship had been physically innocent, and whether it was because of this or in spite of this, I was deeply in love with Valera. With him was the only place I wanted to be for the precious remaining time. We arrived at his family's house in the garden district. His mother had prepared a supper for him, his two sisters, and me. Then she kissed me goodbye and headed off to work. "Mama works nights now, ever since our father died," Valera explained. "He died of alcoholism, but I'm going to follow a different path."

We walked through the garden and looked at the vegetable boxes and grapevine trellises he and his dad had built together. Valera put a spade in one of the garden plots to show me the soil and turned up an old square nail. He picked it up and examined it with great interest. "This must be a very old nail," he said, "maybe from even before Soviet times."

We went into his room and he took down his box of mementos.

He emptied the contents onto the floor, and picked up a picture of his father in a soldier's uniform. He gave me the picture of himself at school, and the red scarf he'd worn as a Soviet pioneer. He gave me his pin of Lenin. He gave me the square nail from his yard. And he gave me my first kiss.

EIGHT

Adventures of an Atheist

My dad stuck his hand into his pocket and pulled out an apple. "Would you like this?" he asked. Taking it, I turned the apple over in my palm to examine its green, waxy skin. Granny Smith. I couldn't find a single wormhole. I took a bite, exposing its stiff, white, spongy flesh. The apple had no taste. I spat the mouthful of nothingness out onto the ground and hurled the apple across the asphalt. I watched it skip and bounce under cars in the Las Vegas airport parking lot.

"In Russia they eat real apples," I told my dad. "They may have wormholes and taste like mildew, but they are real."

My dad didn't speak; he just loaded my green canvas duffel bag, which was stamped with a US ARMY insignia, into his jeep. Stuffed full, it looked like a giant, lumpy sausage. It used to be his. He had gotten it when he was training to be an army medic in Germany. I had been a toddler then, but now the bag was mine and it contained everything I cared about: some clothes, my Russian language notebooks, my journals, and a samovar, a metal urn for boiling tea water.

It was 1993, and I was seventeen years old. I had no idea where my family lived. The last I knew my parents were in Las Vegas, but as I was leaving for Russia, I'd heard they were planning to split up again.

"We all live in Reno now. You and I will have a long drive to get there today," Dad said.

I welcomed the long drive. I was not ready to be home. I was angry that I could not stay in Russia. The American Field Services administrators had said I must to return to the US and then find my own way back. With little money to my name, I knew that meant I first needed to finish high school and get another job. Thoroughly absorbed in every minute of Russian life before boarding the plane in Krasnodar, I hadn't given any thought to where I'd live or what I'd be doing now that I was back in the United States.

I looked out the window as we sped through the brown desert. Once or twice I rolled down the window, yearning for fresh air. The thick, familiar scent of sunbaked sagebrush filled my nostrils, but blasts of hot, dry air were hard to breath. I quickly rolled the window back up, sealing myself inside the quiet, air-conditioned shell of the jeep with my dad.

Finally, I broke the silence. "Where do you work now, Dad?" I asked.

"I'm still with Bally Systems," he answered, referring to the casino gaming company he'd worked for since finishing college in Eugene. "I enjoy the challenges of computer programming, and Bally granted my request to move back to their Reno location."

"Oh," I answered, not knowing what to ask next. Dad had had computers around him long before I ever learned to use one, but despite his attempts to describe how they worked, they always seemed like a mystery to me.

"How was your time in Russia?" Dad asked.

"Good," I answered. "I wish I was still there."

When I was first in Russia, I felt homesick. I would have given anything to be back with my dad, hiking high in the Sierra Nevada.

Now that I was home, I couldn't talk with him. I saw the growing distance between me and my family. What I thought was home, what I had initially longed for six thousand miles away, had never felt farther from me in my life. I closed my eyes and searched my heart. I could hear it beating double time to the cadence of two sets of legs running alongside the Kuban River.

We drove along in silence for a while, but neither of us was the type to keep quiet for long. Dad spoke first. "Katey, while you were gone, I filed for divorce. Hanna and Christa are living with your mother right now. When we get to Reno, you can stay with Annie and me until school starts."

"I knew your guys' remarriage wasn't going to work out," I said, giving microphone to what seemed obvious and drowning out the voices in my head that were asking how Mom and my sisters were doing in the aftermath of what must have been another unpleasant separation. The news of their second divorce did not sting me like the first time, when I was ten. The walls of my heart were tougher now.

Toward late afternoon, Highway US-95 crested a desert hill. The eastern flanks of the Sierra Nevada loomed above us on our left. In front of us, desert scrubland sloped gently downward, terminating in an immense, topaz-blue-colored lake with a string of islands protruding toward the middle. "That's Mono Lake," Dad said. "Would you like to stop for a swim?"

My back peeled lose from the seat as I sat forward in anticipation. Dad sped past the paved turnoff to the visitor center and instead, took a right several miles north onto an unmarked and unmonitored dirt road. We bumped along in the jeep over dry, rocky streambeds. Pungent smells of desert sagebrush flowed into my nostrils. Tall willows surrounded us. I couldn't see the lake, but I knew we were headed in the right direction. When it came back into view, Dad pulled over and we both leapt out.

Stripping off our clothes, we left them in heaps on the seats

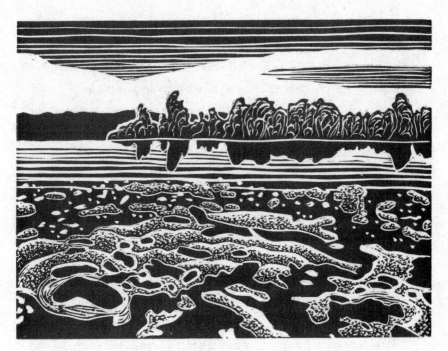

Mono Lake

and dashed toward the water in our underwear and shorts. Dad stopped short of the lake. I kicked off my shoes and sprinted in full force until my feet could no longer find the sand. I closed my eyes and dove beneath the surface, anticipating a smooth coolness that would soon surround me. Suddenly, I was struck by an immense stinging in my eyes. My skin was burning too. Horrified, I stood up. As I did, my toes sunk into a grainy, muddy ooze. I retreated blindly from the water, shrieking in pain. My dad was standing on the shore, still dry in his shorts, chuckling.

"Didn't you know that Mono Lake is saline?" he asked. "It is much saltier than the ocean." For a split second, I thought I was back on the Black Sea, being chided for jumping naively into water before looking, only my dad's voice seemed to hold less affection than Valera's.

I looked around. White crusts of salt covered the muddy shore. Tiny brine shrimp wriggled in the surf.

Where my dad's softness fell short, his inclination to teach me surfaced. "River water enters the lake carrying salt with it," he explained. "But the lake has no outlet. So water just evaporates off the surface, leaving the salt behind to accumulate."

I loved that he was explaining science to me again. "What are those?" I asked, pointing to twelve-foot-tall scraggly spires along the shore. Other spires protruded from the lake itself. They looked like giant versions of the mud-drip castles my sisters and I had made in our sandbox when we were little, except these had turned to fragile, crumbly rock.

"Those are called tufa towers," Dad said. "They form when spring water interacts with the alkaline lake water, causing calcium carbonate to crystalize.

"You *can* actually swim in this lake. It is supposed to make you feel buoyant. You just have to keep the water out of your eyes." With that, Dad took off his shoes and entered the lake. He did a few breaststrokes, keeping his head above the water, then turned over onto his back to call out to me. "Come on, Katey. It's great. You really should try it."

A crust thicker than the coating of dried mud beneath my feet added to the growing callus on my teenage heart. I didn't trust my family. Deciding to close myself off from the only warmth my dad now offered, I sat down on a patch of salt grass in the white mud and hugged my knees.

...

THAT SUMMER DRAGGED on. The letters that came at first weekly from Valera started only to trickle in. I worked hard and earned my first large sums of money as a pasta cook and caterer at the Olive

Garden. But rather than spending a cent, I stashed my money away into savings so that I could one day return to Russia. In the meantime, my pleasure was escaping to the mountains. One morning my sister Annie and I rode our bikes to the end of the country lane where our dad was renting a small shack. Here the Sierra Nevada foothills started to rise out of the valley. We climbed through a cattle fence and started picking our way astutely up a hill, through sandy soil and sagebrush, fearfully aware that we were in thick rattlesnake territory. Reaching the shade of a pine tree on top of one of the sagebrush-covered foothills, we sat down to rest.

We both knew our time together was limited. My family had become to me like lily pads in the pond of my life. I stopped in at home only long enough to learn that the pads were not stable. They could not hold me and so I leapt on. In several weeks I'd be leaving for Portland, Oregon, to live with my dad's half sister and finish high school on my own in a place where I could start taking college courses in Russian at the same time. From there I'd wander farther yet, to the ivy-covered walls of a liberal women's college in Western Massachusetts.

"Katey, do you ever have doubts about God?" Annie asked. "Do you ever wonder if he is not real?" Her questions put into words thoughts I'd struggled with for a long time.

"I think the part of us that believes," I answered, "comes from being told God is real as little children. We were always taught to say we love him." I turned to look into my sister's pretty, blue-green eyes. "But can you honestly say you love him?"

"I don't know," Annie said. "I keep thinking about the people who stand up in church and say how thankful they are that God rescued them from the pit of their sins. They seem like they really feel love for God."

"Yeah, and we've never done anything really bad. We've been little Goody Two-shoes all our lives," I admitted, partly disgusted with myself. I pondered an experiment. "Maybe we need to do some really bad sins and then see if we come back out the other side."

"Maybe so," Annie answered.

We chuckled a little, but as we made our way back down the hillside, the crust around my heart hardened further. With every step down the mountain, I reconciled for myself a new path forward. Sure, I had shared something special with Valera in Russia, but hadn't a good part of that been just a crush? His most recent letters indicated he'd left Krasnodar and was living on the road with dance troupes performing around Russia and Europe. They were on their way to Turkey. It was hard for me to imagine in that environment he could be true to his faith. My own faith, at the times it was alive as a teenager, attending church both in Eugene and in Krasnodar, had been real, but it was conflicted. One part of me believed in God and another part questioned the reality of it all. As I descended the mountain, I forced myself to dismiss the light I had seen in the Russian Christians. It was time to act on my experiment. If I still had doubts after all of these years of living as if I were a good Christian girl, never committing any blatant sins, like having sex, lying, and taking drugs, then it was time to start living like I did not believe in God. This didn't mean I was going to let any new worldly activities derail me from an immaculate educational record in the US, which I believed was the linchpin to my future, but I decided to turn my back on God and the rules of Christianity. From that moment forward I was going to live like there was no God.

...

IN 1994, as an eighteen-year-old freshman at Mount Holyoke College, my only religion was school. I put my full faith and efforts into my college education. I counted on it alone to someday provide me with security. I could afford this private liberal arts college only through scholarships and a part-time summer job I had at a local restaurant. I was desperately fearful of debt and spending money of any kind. Not knowing where my basic necessities would come from and at the same time thrilled by the possibilities presented once I'd abandoned my Christian morals, I soon found myself in a vulnerable position. I was in a secretive romantic relationship, in which I was dependent on someone else for my basic needs. I was excited and scared at the same time, and I had written God out of the equation. Days and nights were consumed with study. My personal life, my hidden life, was a narrow margin in which right and wrong no longer existed. Guided by an ignorant, self-centered instinct for survival, my conscience slipped into darkness, a darkness I dared not share even with my closest friends. One rainy day toward the end of my sophomore year, I lay in bed looking at a mosaic of large, green maple leaves pressed against the window glass in my Western Massachusetts college town. Water streamed down the glass and across the leaves. I was acutely aware that if I did not change my mind toward God, my mind was going to split. I would go insane. Water means life, and the rain battering my window revealed an insurgence, though tiny at the time, of faith.

I checked out the campus Christian society and found the girls knitting there to be a bore. I wondered if there was any hope in finding God in my curriculum. Although I had declared a major in geology, I took as many biology and ecology classes as I could from Stan Rachootin. Stan had an infectious passion for the wonders of the natural world, and when I asked him one day in the privacy of his office if he thought the living world around us allowed for God as a creator, he replied in the absolute negative. I had heard rumor that another professor on campus, a paleontologist, held a different opinion. I was drawn to learn the perspectives of both of these men, but I had never personally met the paleontologist, and Stan's remarks were persuasive enough that at this stage in my life, I did not seek further.

Desperate to escape the suffocating hole of my private life at this women's college and hungry for more adventure overseas, I found a way in 1996 to apply my scholarship money to study geology during my junior year in Edinburgh, Scotland. In Scotland, I discovered that my interest in the natural world would never be satisfied by studying a single discipline in isolation. I knew geology was important, but it was only one piece of the earth-system puzzle. I would need to learn more about biology, ecology, chemistry, and the interactions of all of these disciplines.

My year away also revealed to me how much I appreciated the real world where men and women coexisted. There were some women at Mount Holyoke who professed to feel more comfortable in an all-female class setting, but I could not relate to this. I enjoyed my female classmates, but I'd never felt intimidated by males before and didn't see the point in creating an unnatural all-female society. Besides, Scotland was a lot closer to my dream of finding Valera. It had been over three years since I'd left Krasnodar. I'd seen Valera briefly in Krasnodar in 1995 on my way to Novosibirsk. That visit confirmed my desire for him and the purity he had once represented in my life. I felt that I simply must see him again. Perhaps if I saw him face-to-face I could find innocent love and spiritual light again.

I worked hard enough in my geology and ecology courses at the University of Edinburgh to bypass midyear exams. This gave me extra time for winter holiday travel. Following a clue from the latest letter I'd received from Valera more than six months prior, I bought a cheap round-trip ticket to Istanbul and decided to try my luck at finding him in Turkey. I had three weeks and no lack of determination. How many Russian dance troupes from Krasnodar could there be in Turkey anyway? The Russians had a saying that "by your tongue you will reach Kiev." I would use my tongue to find Valera. After all, my elementary school motto had been "Where there is a will, there is a way," and I still lived by it.

Valera's last letter had said he was in Izmir. I found Izmir on a map—it was a coastal town situated on the Aegean Sea. I bought a bus ticket there but soon found myself the only female on an overnight bus. The trip was agonizing, particularly as I could see no way to escape my situation. On the back seat of the bus, I pretended to sleep, but I was terrified of the men who took turns advancing into my personal space and stroking my legs. When the bus finally pulled into the Izmir terminal, I didn't care a bit that it was still dark outside. I wanted to get off that bus as quickly as I could. I slung my pack onto my back and took leave of the station in as big and strong and unattractive form as I could muster. I strode away feigning confidence that I knew where I was going—I didn't want to give the men lingering around the station any impression of my true fear. I was tired, but I walked until the sun rose in the east and I heard the first call to Muslim prayer blare through the city's loudspeakers.

I walked until I found the seacoast and the tall tourist hotels that lined its beaches. Now I knew I needed rest. Since I could not afford a fancy hotel, I found a hostel several streets away, paid two US dollars for a room and breakfast, and collapsed onto the large square pillow at the top of my bed. When I awoke, I was directed to the roof of the hostel, where I sat beneath a reed-woven canopy, and waited for my breakfast. Again I heard the call to prayer, and I watched as the hostel staff gracefully disappeared from view.

The sun was well up. I peered over the side of the flat concrete rooftop. Light smog shrouded a cluster of minarets towering above a sea of other gray and red-tiled roofs across the city. Some people were bowed on rugs, praying in a neighboring courtyard. Minutes later my breakfast arrived; I resisted the urge to say grace. When I had finished eating, I chose a grand-looking hotel along the beach's promenade and walked up to the front desk to inquire if they'd heard anything about the whereabouts of a Russian dance troupe. "A lot of dance troupes come through here," the man said. "But I remember one from Krasnodar here about two months ago. You can look in some other towns, maybe Marmaris, Kas, or Antalya."

I took buses to each of these towns in turn, searching the faces and imploring the whereabouts of the Krasnodar dancers. My time was running out. With only five days to spare, I arrived toward evening in Antalya. In the foyer of one large hotel, the director of tourism, sharply dressed in a business suit, raised his black eyebrows at my inquiry. "Yes," he said. "We do have Russian dancers here. Two males and six females. They will perform tonight."

"Could you please tell me, is one of the male dancers named Valera Fedoseev?" I implored.

"I would need to inquire," the man answered. "Please wait here."

My heart pounded as I took a seat in the high-ceilinged, spacious room. I must have found him! Would Valera himself come out to meet me? I wondered what I looked like. Would I be pleasing to him?

After what seemed ages, a petite, prematurely aged woman with thick makeup and short, curly red hair walked briskly across the marble floor to me. "Hello," she said in English, nodding. "Who is it you are seeking?"

"I'm looking for Valera Fedoseev," I answered her in Russian. "He is an old friend of mine. I knew him when I lived in Krasnodar."

"*Da*, Valera Fedoseev is here. He is getting ready for a concert. You can meet him tonight. After the concert."

He was there! I had found him at last! Knowing this was enough for me, but I was disappointed that the director did not invite me to watch the concert. The hours passed painfully slowly. I paced back and forth in the foyer, watching the sky grow dark outside and night fall. Several times I visited the bathroom to straighten my clothes, check my hair, and apply a bit of lipstick. The lipstick was something I did only on very special occasions. I tried reading the book I'd brought along, but it was too hard to focus. I wondered what it would be like to see Valera again. Was his director going to tell him an American girl was here to see him, or would I be a surprise?

Finally, at the moment when my giddiness was fading into drowsiness, I heard a shuffle of feet. Valera entered from stage left, still in costume. His blond hair was short now. He wore tight pants and a muscle shirt. His jaw dropped open and his makeup-covered lips parted in a huge smile, showing the same mouth of spaced teeth I remembered from our youth. "Haaaa! Katushka!" Valera shrieked in a high-pitched tone. He leapt across the marble floor, gathered me in a warm embrace, and planted a big kiss on my forehead. I couldn't believe this was real. It had been such a long time since I'd spoken Russian. The words that came out of my mouth were stiff and awkward to form. I stumbled both in body and tongue as I tried to express my gladness at seeing him once again too. We stepped back to look at each other; my smile was so big it hurt.

Valera took my hand and led me to the dressing room to gather his things. There he introduced me to a Black Russian who was his male dance partner and to the ensemble of female dancers. I'd never seen girls so skinny or so beautifully made up before. Sharp black lines accentuated their brows and eyes; their lips were painted dark red. Luscious hair fell down their backs, covering more of their bodies than the costumes they'd worn on stage. I was shocked at how little body hair these girls had, and even more so that I could see enough of their naked bodies to notice. One girl with long brown hair and a particularly warm personality came over and welcomed me. She seemed sincerely glad to meet me. Not all the girls were there, she explained. I asked Valera where the other girls were, and he explained, in a somewhat subtle way, that they were expected to be available to make money prostituting on the side after the show.

We left the plush dressing room and exited out a back door of the hotel. Instead of going up to luxury suites, which I'd expected all along, Valera led me to a ramshackle dormitory in the hotel's maintenance yard. Large plastic bags of rice and crackers lay on the floor next to a small propane stove. On the counter stood a tea boiler, a bowl of sugar, and a plate of dried, salted fish. These dancers were not rich! Hidden behind their beautiful costumes, they were the poorest of poor. Valera did not apologize or show the least bit of shame. He was happy to see me.

I was happy to see him too. After three and a half years of living on blissful memories of our time together as teenagers in Krasnodar, I was finally here with Valera. But sitting there in the dancers' tight living quarters with my large bones filling my jeans and sweater, I felt insecure and out of place. I was disappointed to see that Valera's life had deviated so much from the faith we once had shared in God. It felt like a rejection of the light that was our relationship. *That's okay*, I tried to justify it to myself. After all, hadn't I turned my own back on God?

"Would you like to eat?" Valera asked. We ate salty dried fish and crackers and drank cups of black tea with sugar. Then he provided me a place where I could sleep just as the sun started to rise above the sea.

Valera filled the next three days exposing to me the best of his life. In his off hours, we swam in the hotel's pool, ran barefoot on the sandy beach, and ate fresh seafood and exotic fruits in beachside cafés. Valera's smile never faded. He was excited about his opportunities and accomplishments. The stunts he'd performed on the rusty playground years ago by the Kuban River had since transformed into professional dance moves. He was proud to have me sit in the theater, a lone spectator, watching his troupe practice hour after hour in the afternoons before the nighttime performances. The dances were exotic and sensual. They made human pyramids and enacted the taming of wild animals in a jungle. Although their frames were far slighter, the women were as strong as the men, lifting and swinging one another through their legs and through the air. I was never permitted to see a nighttime performance. Those were for paying customers only. But I was content to rest and wait in the dorm with a bag of crackers and my book.

When the concert ended on my final evening in Antalya, Valera took me out on the town to a beachside disco. We had a few drinks, and he shouted to me above the noise of the crowd, "*Pashli tansivat* (Let's go dance)." I had been dreading this invitation. I did my best to move with the people and the music, but Valera quickly laughed and said, "You still dance like a bear." I knew with my large build and lack of rhythm that I was no competition for the girls he was used to dancing with. In fact, I was so far out of their league that I said to myself, *If Valera is going to love me, it will have to be for who I am and not for how I dance.* But his words stung. I could not argue with them—I was sure that if anyone was qualified to judge, it was Valera. I danced like a bear. And maybe that mattered. Clearly, our lives were different and we were different. Still, first love dies hard, and I left with the fading hope that the purity of our first love would someday be rekindled.

...

I MAY NOT be able to dance, but I will strive for excellence in science, I resolved as I made my way back to Scotland and then through my graduation from college. In the spring of 1998, I accepted an offer from the limnologist Dr. Charles Goldman to start graduate school at the University of California, Davis, studying the immense, pristine lakes Baikal and Tahoe. Going to graduate school in ecology meant that I had to turn down a well-paying job offer from a New York City–based oil company with operations in Russia. But often it is the difficult decisions we make at life's crossroads that require us to search our own souls and discover who we really are. I was destined for lakes. The life of a graduate student wouldn't make me rich, but I didn't care as much about money anymore. As long as I had enough to eat and pay rent, what mattered to me now was becoming a real scientist, turning my love for lakes into a deeper scientific and professional knowledge of them. Besides, UC Davis was just on the other side of the Sierra Nevada from my family in Reno. I'd seldom seen them during the past eight years, and the longing for a sense of home drew me back to them.

I bought an airplane ticket to Reno, but rejection met me at my dad's front door. While I'd been away, he had remarried. He was now living with his new wife, Sandy, in her house, and they made it clear that their home was not my home. I was welcome to visit them, but only for pre-arranged meals and gatherings. My mother welcomed me with open arms, but she was sleeping at her own parents' house in Carson City and didn't really have a home to offer. It became evident that I must trudge onward along the path I had set out on at age twelve, when I had first left my mother and sisters to live with a family from church in Eugene.

To get to my new graduate school position at UC Davis, I would need a car. Dad offered to sell Sandy's little red Honda to me at a low interest rate. When I asked why they were selling it, he said, "I want Sandra to have airbags. Her Honda doesn't have any. We're going to get her a new car."

"Don't you want me to have airbags too?" I asked.

"You can have airbags when you can afford airbags," my Dad replied, his dark eyes looking intently at me past his pointy nose.

Tears welled in my eyes. Behind the blur, childhood memories of begging Dad to buy me gym shoes and underwear returned. He was the only parent who'd ever had any money, but he did not let go of it or of his time for me very easily. I remembered the years of playing competitive sports in high school. I was always a starter on the basketball and volleyball teams, yet it seemed like I was the only player on the team who didn't have a parent willing to drive me to practices or watch in the bleachers.

Now, as a young adult, he criticized me for being self-centered. He and Sandy didn't appreciate it when I arrived late to an invited meal or canceled plans at the last minute.

"I feel like you don't love me enough to care for me, Dad," I said, crying angrily one afternoon at his front door.

"If you feel that way, then we don't need to have a relationship," he said coldly with anger rising in his own voice. "To be welcome at our home, you need to become more considerate first."

I left brokenhearted. I did want a relationship. That is what I wanted more than anything, but it seemed that I was once again left to my own devices.

...

IN CHILDHOOD, I had gawked at the brilliant blue water of Lake Tahoe. When I had looked from three hundred feet up, on the lake's rim, huge white boulders were visible tens of feet below the water's surface. But at UC Davis, Dr. Charles Goldman, my new graduate program advisor, told me that a noxious weed, known as Eurasian watermilfoil, had invaded Lake Tahoe and might be aiding the lake's water quality problem. When scientists began measuring Lake Tahoe's clarity in the late 1960s, the dinner-plate-style secchi disk could be seen down to one hundred feet. When I arrived to start my master's degree program in 1998, the disk disappeared from view at only seventy feet. Each year it was getting worse. The culprit was microscopic algae and other fine particles that scatter light.

Lake Tahoe

Algae grow thicker when fed phosphorus. Phosphorus was entering the lake water from somewhere, turning the lake green. But where exactly this phosphorus was coming from, nobody knew. It was possible, I thought, that Eurasian watermilfoil weeds were helping to supply the algae with phosphorus.

I wanted to help solve this problem but didn't know where to start. I needed a mentor but didn't have one. Dr. Goldman ran a huge research program. He was kind and encouraging to me personally, but he did not have time to train me in the details. Desperately wanting not to waste precious time and scholarship money, I called my Grandpa Sid for sympathy. "Follow your intuition, Katey," he said. "Intuition is what led me to my best real estate transactions." So I took classes in weed science and bought a wet suit. My intuition said I should get into the water and have a firsthand look at the problem.

I learned that Eurasian watermilfoil grows entirely underwater. It roots itself in lake sediments and sends tall stems with whorls of feathery leaflets up into the water column. I thought it was ironic that the subject of my study was the very thing that often caused me to panic while swimming. I didn't like the surprise of my feet brushing against slimy beds of plants hidden beneath the water's surface. Here was an opportunity to unwrap that fear. I knew I would need to map the spread of this plant around the lake and conduct careful laboratory experiments to determine if it was indeed responsible for some of the increasing phosphorus in the lake's water.

With deliberate breathing and a waterproof notebook in hand, I dove into each of the major bays of Lake Tahoe, mapping the distribution and density of the Eurasian watermilfoil stands. For the first time in my life, I was swimming with a purpose. I was no longer swimming just to escape the world above the water. Now I was here also to understand the world in the water. To understand it, I needed to watch it and touch it, to move with the streams' currents as they dumped fresh water into the lake's lush deltas teeming with milfoil plants, to feel the sunlight warming my belly through my thick black neoprene wet suit as the rays reflected off the bright sandy lake bottom. Science was the path I was choosing, leaving behind me all thoughts of family and faith.

As I swam out deeper, I saw millions of tiny milfoil fragments floating around on the vast open water of the lake. These fragments grew long white roots during their journey, and when opportunity presented itself, the adventitious roots took hold of sediment along a different beach. They were pilgrims, giving rise to whole new plant colonies. My laboratory experiments using a radioactive tracer, phosphorus-32, showed that the plants were indeed transferring phosphorus from sediments to algae. The spread of Eurasian watermilfoil was contributing to the diminishing water quality in Lake Tahoe. I saw no hope for controlling this strategic weed. The use of chemicals would never be permitted in Tahoe, but it seemed that the mechanical mowers operated by wealthy property owners in marinas were only exacerbating the problem by spitting out even more tiny plant travelers.

Although they were never published, my scientific results were of great interest to California policy makers. I was pleased that my first experiments established a foothold for my scientific career. I was no longer just a wilderness vagabond but had generated useful scientific results. My effort to take a closer look at things lurking in the depths brought me greater clarity and courage when I returned to the surface. Now I was inspired to keep moving forward, to keep looking for answers to bigger questions.

In 2000, I received a US Environmental Protection Agency STAR Fellowship to fund doctorate level research, a turning point in my life that led to my study of permafrost with Sergey. Without my research in Lake Tahoe, this opportunity might never have presented itself. I was torn between continuing to study lakes in the Sierra Nevada I loved and moving to the University of Alaska Fairbanks to study Arctic lakes in Siberia. Relationships held no tethers on this decision. Science itself was becoming my devotion. While I had seen Valera again in Krasnodar in 1998, on my way to spend a summer working on environmental problems at Lake Baikal, it was clear the pursuits of our lives had diverged. He was a European dancer, and I was an American scientist. The innocent love we'd shared as teenagers was a dim light that still attracted me, but staying in California or moving to Alaska bore no weight on any future Valera and I might share. This chapter of my life was my own.

Somehow I knew it was my calling to relocate to the Arctic wilderness and conduct research in my element: water. I felt a tinge of guilt for my selfish decision to walk away from the human-made environmental problems at Lake Tahoe, but I trusted my peers' abilities to address them without me. Edward Abbey's advice called to me loud and clear: "It is not enough to fight for the land; it is even more important to enjoy it. While you can. While it's still here." The adventure of studying the Arctic wilderness, far away from problems caused by humans, beckoned to me. And it would pay off: a few months after making this choice, I would end up studying methane bubbles under Terry's gentle mentorship on the University of Alaska's Fairbanks campus and Sergey's gruff yet visionary leadership in Cherskii. But what I did not know at the time was that even in the most remote reaches of the vast Siberian tundra, my scientific research would not be isolated from human impact. Hovering over the Arctic, a halo of fossil fuel emissions was causing permafrost to thaw.

Part III

COMING OF AGE

NINE

A Scientific Author

At first meeting, a lake will reveal only so much about itself. Usually, it will be the things I see in summer, when I too have shed my wintry outer layers and stand momentarily exposed at the water's edge. My toes may burrow into soft beach sand. I may tower on boulders or peer with caution over a steep rocky cliff. Thick trees and shrubs may crowd the margins, permitting me to peek at the lake's watery surface only if I dare to balance on a log that fell across its inlet creek. A lake may be an oasis in a sea of prairie grass, flaunting its outward appearance from every angle. Some lakes don't let me get near them at all—turquoise lakes perched in the crotch of glaciated mountain peaks with nothing along their edges but steep inclines funneling talus. Other lakes are not so intimidating. A small, sheltered lake rarely gets so much as a ripple across its surface. A big lake is quite the opposite: it is a racetrack for the wind, where breaking waves applaud the performance. What I love about all lakes is that they always win at playing hard to get. I may put on a snorkel or scuba-diving suit, but since I am not a copepod or a fish, I am not allowed to get a good intimate look at what is inside a lake. In this way a lake guards its mysteries. It flirts with me and tempts my curiosity.

Back at my little cabin on Chena Ridge in November 2004, I watched through the window the ice fog in the valley below. I thought about the path that had brought me to this moment. I'd spent my adolescence running away from—or looking for (I'm not sure which)—home, and I found science instead. I had lived the first twenty-four years of my life on the margin of a temptation to understand lakes. In the beginning, lakes had been fun places to splash and play. As a teenager, I escaped to them. Their beauty caused me to forget what I thought at the time was ugly in myself and in the world around me. When I asked my dad if it was possible to stay at the lakes and make a living there when I became an adult, he told me that I should not make a career out of something I enjoy. Turning it into work might take away the joy. I believed him, at first. So in college I thought maybe I should become a medical doctor. That was a respectable, financially secure career. But I could not help myself. Lakes always drew me back. By the time I reached Cherskii in 2001 to study methane emissions with Sergey, I had turned a corner in my life because I knew it would always be centered on lakes. From the first summer fighting muskrats to the final breakthrough as I singed my eyebrows on the frozen lake, I had found a way to work on the same thing I loved.

Now, on this autumn morning, I was knee-deep in my doctoral program. I moved from the window to the kitchenette to make myself a cup of tea, then climbed into the loft to continue working. The heat from the woodstove was dying down, but I would remain warm for a few hours yet. I found myself frustrated: my doctoral research had so far been an enormous turning point for me as a scientist and as a person. Four years into it, I was having my first breakthroughs, discovering things in lakes that would change how science understood climate change. I was no longer escaping to lakes; I belonged there. I was starting to make them my home.

However, I was at a loss as to how to communicate to the world the discovery of ice bubbles Sergey and I had made. I needed my advisor, Terry. A few days later, he stepped into the abyss of my office in the Arctic Health Research Building on the UAF campus. Stacks of papers loomed overhead. Dried out tea bags hung off the edge of my mug. Terry sat at my desk with me for thirty minutes. I presented him with the muddled, blurry pile of graphs and data tables I'd been compiling. How should I begin to write up my results? How should I tell my story? To which journal should I submit my first paper? I'd been swimming circles in this puddle of questions for weeks, anxious, blind, and confused.

Terry stirred the murk for a few moments, asking me some pointed questions. Then he said, "I think you could submit your work to *Nature*." He had spoken calmly, but his words fanned a spark of gainful ambition in my heart. *Nature*! That was the most prestigious of all scientific journals. Terry thought my first paper had a shot at *Nature*! Maybe all of my efforts were going to pay off after all. Maybe I would stop being a scientific nobody and become an author, even an author with a name in *Nature*.

After twenty-four iterations of my manuscript and great editorial input from Terry, I finally had a story to publish. Using my new methods, I was able to prove that there was a lot more methane escaping from Siberian lakes than anyone had previously thought— five times more! Siberian lakes would now be rivals with northern peat bogs and wetlands, environments that have long been known to release the potent greenhouse gas. But unlike wetlands, which mostly recycle modern carbon from plants, Siberian lakes were belching out ancient carbon that had been locked away in frozen Ice Age permafrost deposits for thousands of years. As the permafrost thawed beneath the lakes, microbes were feeding on the ancient soil carbon and producing methane. Gas bubbles wobbled up from the sediments to the lake's surface, where they popped, releasing the methane into the atmosphere, where it was lifted away by air currents and carried around the globe, contributing to the greenhouse effect worldwide. Global warming was felt there in Siberia too—warming caused more ice wedges to thaw and more methane-belching lakes to form. What I had discovered, together with Sergey and Terry, was a positive feedback cycle to climate change.

With one manuscript on its way to *Nature*,[1] I felt empowered to write a second paper to the other top research journal, *Science*. This time I didn't want to rely on the crutch of my advisors. I needed to start proving to myself that I could come up with important questions without them. After all, someday perhaps I would run my own research lab, and then it would be up to me alone to navigate the sea of scientific inquiry.

The idea for my second paper came about after a discussion in front of a conference poster with a new colleague, a gregarious paleoecologist (a person who studies ancient ecosystems) with a generation more than me of experience. She had stopped at my poster because the thermokarst lake methane bubbles had caught her eye. As we talked, our unique backgrounds gave rise to an exciting new question: Had lakes been an important feedback to climate warming in the past, when Earth emerged from the last Ice Age? This colleague told me that polar ice core records from Greenland and Antarctica showed there had been an unresolved source of atmospheric methane in the Northern Hemisphere at the time when Earth started to transition out of the Ice Age. Perhaps Siberian lakes were the mysterious source! Perhaps they had formed when the climate started warming back then, accelerating the warming as they appeared to be doing now too. Electrified by this idea, I set in to compile the records and refine the ideas with this new colleague and Sergey, and soon I had a second manuscript on its way to *Science*.[2]

With two such high-profile papers, my fears about being a nobody, rooted in feeling as if I were not good enough for the big, affluent world beyond my financially challenged and unstable childhood, started to dissolve. If I could get these papers published at *Nature* and *Science*, surely I'd have a chance for recognition and success in academia.

In the upstairs alcove of my cabin on Chena Ridge, I took time in between editing my manuscripts to sprawl on the loft floor next to the window that looked out over the birch forest canopy to the silhouette of Denali on the horizon. As my scientific potential seemed to be escalating, I realized I needed to give time to the empty places in my heart and soul too.

It was early on a Friday evening in December 2004. My cabin was quiet and I was lonely. I thought back to the events of the week, which were the culmination of four years of bouncing between Russia and Alaska. I had been self-centered in romantic relationships, and I had never learned to love others or myself well. The result of my frivolity was simultaneous engagement prospects from either side of the Bering Strait. I admired Dima, my Siberian field assistant, but I didn't see marrying him as a practical future. In Alaska, the university forestry student I'd been dating for several years seemed a more plausible match, but I was not ready to get married. One of the obstacles in my mind was that he was not a Christian. I realized this concern was contradictory to the selfish and secular lifestyle I myself was leading, but in the back of my mind was the warning I knew from the Bible, the warning not to be unequally yoked. This meant that a believer in Christ should not marry an unbeliever. Feeling a little crazy in the awareness of my own hypocrisy, since I had not committed to calling myself a believer, I went to seek professional consultation.

Four days had passed since Monday, when I had walked unannounced into a church on a street corner next to the university. The man I had found in the office labeled PASTOR looked up from his desk somewhat startled. When I asked if we could talk, he sat back in his chair, relaxed, and was ready to listen. I felt safe confessing my fears and concerns since I had no ties to this man and would probably never see him again. My first fear was that I might be crazy. He assured me I was not. My second fear was about disobeying the Bible's instruction not to marry an unbeliever. The pastor told me I didn't have anything to worry about with that either. "God works all things out," he said.

I walked out of that church feeling better, but something bothered me about the religious leader's dismissal of scripture. I went home that Monday night and opened my Bible. It was the same soft, gray, leather-bound book I had carried to Russia during my first trip there. I hadn't opened it much in all the years before moving to my cabin, not until God had answered my prayer for victory in the muskrat battle. But in the past few years, I had opened it from time to time. On that Monday night, I decided to ask God himself if he had anything to say about my dilemma.

Digging my thumbs into a random place somewhere in the middle of my Bible, I opened to: "They did not wait for his counsel, but lusted exceedingly in the wilderness, and tested God in the desert. And he gave them their request, but sent leanness into their soul" (Ps. 106:13–15, NKJV). I knew that these years of loneliness in Alaska and Siberia were my wilderness; they were my desert. This scripture was God's direct warning to me. If I married the forestry student, my request for companionship would be granted, but the cost would be to my soul. My soul felt lean as it was, and it became clear to me in that moment that what I wanted more than anything else was health in my spirit. And so, declining marriage proposals, I found myself once again alone, at least in a physical sense.

Now it was Friday. My cabin was particularly still. I didn't know where to go to escape the pain of loneliness. Even the crackle of wood in the fire was not a consolation to the emptiness that surrounded me. I realized that if I was giving up tangible human companionship for a God I wasn't sure I believed in, I needed to get a few things straight inside myself. Since my dad was the initial reason I'd ever doubted God in the first place, I thought that talking with him might help me get to the root of the matter. Mustering up some courage, I curled up on my futon couch next to the woodstove and phoned my father in Nevada.

"Dad," I said, after giving him an update on the successful progress of my studies, "what I really want to talk with you about is the confusion in my personal life." I gave him an earful about my decisions not to get married and the tormenting contradictions swirling inside me. Fully expecting the same derision of any notion of God he'd always given me, his response caught me off guard. His tone was predictably critical, but the message was different. "Katey, you've got to decide one way or another. Stop walking the tightrope of doubt. Believe or don't believe, but choose one."

Later, after putting down the receiver, I sat on my couch dumbfounded. For the first time in my life, my dad had left open the door for me to believe in God. It felt like a chain had been loosed. I was no longer bound to my dad's expectation that the correct and rational decision was a world without God. Unbound, it was now up to me to decide if I would believe in God. Did I dare take a step forward toward faith? I knew this was not something I could do on my own. The doubts in my mind were too big. If God was good and the creator of all things, then where did evil come from? Was Jesus Christ just a historical figure, or was he really the son of God who performed fantastical miracles, like rising from the dead? Over the ensuing months, I read books on Christian apologetics[3] and found that they agreed with C. S. Lewis's statement that, based on Jesus's own words, he could not be just a good man, teacher, or prophet. Lewis implored, "I am trying here to prevent anyone saying the really foolish thing that people often say about Him: 'I'm ready to accept Jesus as a great moral teacher, but I don't accept His claim to be God.' That is the one thing we must not say. A man who was merely a man and said the sort of things Jesus said would not be a great moral teacher. He would either be a lunatic—on the level with the man who says he is a poached egg—or else he would be the Devil of Hell. You must make your choice. Either this man was, and is, the Son of God: or else a madman or something worse. You can shut Him up for a fool, you can spit at Him and kill Him as a demon; or you can fall at His feet and call Him Lord and God. But let us not come with any patronizing nonsense about His being a great human teacher. He has not left that open to us. He did not intend to."[4]

For me, the evidence pointed to only one conclusion: Jesus was who he had said he was, God! This evidence appealed to my mind, but it was the door of my heart that needed to be moved, to be opened. After nearly a decade of concerted effort to dismiss the concept of God, I could no longer ignore the quiet knocking at my heart. I opened my Bible again, but first I prayed: "God, I feel like I have one foot in a bucket of yearning for belief in you, but my other foot is stuck in a bucket of doubt. If you want me to fully believe in you, then you will have to give me faith."

I read Romans 1:20–28 (NKJV): "For since the creation of the world his invisible attributes are clearly seen, being understood by the things that are made, even his eternal power and deity, so that they are without excuse, because, although they knew God, they did not glorify him as God, nor were thankful, but became futile in their thoughts, and their foolish hearts were darkened. Professing to be wise, they became fools and changed the glory of the incorruptible God into an image made like corruptible man . . . And even as they did not like to retain God in their knowledge, God gave them over to a debased mind, to do those things which are not fitting."

This passage described my life for the past twelve years. When I turned my back on God, a darkness had come into my life. But for some reason he had not yet completely cut me off. Inside of me was still a desire for the light, for what was pure and good, and to be right with God.

In John 8:12 (NKJV), Jesus said, "I am the light of the world. He who follows me shall not walk in darkness, but have the light of life." In John 10:10 (NKJV), he said, "I have come that they may have life, and that they may have it more abundantly." I wanted this light and I wanted to live with an inner abundance, but I wondered what believing in God would mean for my aspirations of becoming a scientist. After all, most scientists were atheists, weren't they? Didn't they, like my dad, believe Christianity was just another fairy tale? Would my chance at finding a way in the world and fulfilling my dream to become a professional researcher be undermined by a reputation of being a Christian? I was realizing this was a risk I would have to take.

I started to stake my heart on promises I found in the Bible. "For he [the Lord] satisfies the longing soul, and fills the hungry soul with goodness" (Ps. 107:9, NKJV). "Delight yourself in the Lord and he shall give you the desires of your heart" (Ps. 37:4, NKJV). Like a fragrant oil lathering the dry, cracked state of my soul, these words soothed my innermost being. Truth spoke to me in those pages. I would not close the cover on truth again. I left my gray leather Bible open on the floor and went back to work.

TEN

Truth Pursuit

had discovered the importance of methane bubbles in Siberian lakes. But I knew my work was not finished. I had a sense that it was only beginning. How could I be satisfied knowing about the bubbles in a handful of lakes near Sergey's science station? What about the rest of the millions of lakes in the Arctic? Were they not also surrounded by permafrost? I needed to know if ice-trapped methane bubbles existed in other Arctic lakes too.

In the fall of 2004 I had persuaded fellow graduate and undergraduate students and even friends to come out and stand on the shores of unfamiliar lakes in Alaska, holding a rescue rope should I fall through dangerously thin ice. I realized that the fall freeze-up of lakes was my precious, narrow window of opportunity to map the bubbles and set my traps before thick snow or frozen slush masked the secret of the bubbles' locations for the rest of the winter. As soon as ice formed on the lakes in Fairbanks, I left the writing of my dissertation in my Chena Ridge cabin loft and went out to the lakes with broom, shovel, notebook, and bubble traps in hand. I performed ice-bubble work on as many lakes as I could around Fairbanks. Then I drove 360 miles up the Dalton Highway to the myriad of tundra lakes surrounding Toolik Field Station in the northern foothills of the Brooks Range. All of this effort revealed to me that a plethora of methane bubbles were in Alaskan lakes too, but I saw that the regions were different. Lakes in some regions were releasing more methane than lakes in other regions.

Methane bubbles in the Brooks Range

Nearly a year later, the fall of 2005 was approaching and so was my PhD dissertation defense date. Unexpectedly, I received an offer from an established professor at the Woods Hole Oceanographic Institution in Massachusetts to go to a different part of Siberia. This was an opportunity to conduct my ice-bubble fieldwork on lakes along the Lena River north of a little village called Zhigansk. This professor, who studied river hydrology, had money remaining on a grant but could not go to the field himself. If I agreed to collect water samples for him along the Lena River, then I could also take time to study lake bubbles there. I knew that going to the Lena River would delay my dissertation defense, but it was a chance to go to a part of the world where very little scientific data of any kind had ever been collected.

I struggled with the decision. Scientifically, I knew there was much to gain. Personally, what did I have to lose? Only a little further delay in my chances of finding a husband? I was twenty-nine years old. All of my best friends were married, and I longed to be married too. I longed for the security and coziness of coming home to be with another person at the end of the day instead of to my empty cabin. I thought back over the events of the past year. My deeply buried hope that one day Valera and I could rekindle our childhood love had been a factor in turning down other marriage proposals. But this hope was now dead. It had died in the spring.

Earlier that year, our lifelong correspondence, which had been based on an occasional exchange of letters, suddenly changed to several months of frequent phone calls. If we were to get married, Valera said, he'd be willing to leave his job in the little Italian coastal town of San Benedetto del Tronto to come live with me in Alaska. Surely he could find work in Alaska too. In May I packed my hopes along with my field and lab notebooks into a carry-on bag (the notebooks were so precious I never risked losing them in checked baggage) and boarded a plane for Italy. My goal was to analyze my data at Valera's kitchen table in San Benedetto while he was at work and get to know him better in person again before taking the giant leap into marriage. Just to be prepared, I bought a wedding gown for a hundred dollars and tucked it into my checked baggage. But in all the five weeks I was in Italy, there was never the need to unpack the dress.

Valera had aged. I didn't mind that he'd filled out to look like a slightly plump middle-aged man. My love was as earnest as it had been when we were teenagers. Valera no longer danced professionally, but he taught at a dance school part-time. And he still had his costumes. The first thing he did once I had put my bags down was don the costumes in sequence and dance, out in the courtyard behind his house. I stood beneath a flame-red bougainvillea vine and watched.

Then Valera went to work. He still worked nights, but now as a waiter at a pizza bar along the beach. He introduced me to his friends so that I would not be alone in the evenings. Most of them were prostitutes, some Russian and some Italian. One of them gave me an amber wristwatch that I wear to this day. It quickly became evident that Valera lived in a world of blurred reality. Many of his friends thought that he and I were already married. They said that according to Valera, he'd gone to Las Vegas and married me the year before. I also heard mixed stories of a girl who was still in a coma in the hospital after a motorcycle accident. Three years ago she had been riding behind Valera when he was high. Valera said he didn't do those kinds of drugs anymore, and according to him, the girl was alive and well. He didn't like being confronted with these mixed stories which I presented him with early one morning when he got in from work, but he never became angry. He only had a sad and distant look when he turned his eyes away and sighed.

Moments later he jumped up from the table and said, with newborn excitement, "Shall we go play tennis? It's almost 5 a.m. The sun will be up soon. We can play before the people who pay arrive at the club!"

"Sure," I replied. We hopped on his moped and sped off toward the courts. They were not officially open yet, so we scaled the fence and dropped down into the powder of the red clay surface. We hit the balls hard until the sun rose above the horizon. With each swing of my racket, I drove the dreams of my childhood away. I was training to be a scientist. Truth was important. It was the ambition of my life. I could not marry someone who no longer knew what truth meant.

The iron hinges on the entry gates creaked open. A group of middle-aged men came out to the bleachers with their sports bags and rackets. Valera and I collected our balls and left. Before reaching the moped, Valera stopped at a lamppost. He put his head and hands on the post and kicked his legs up into the air. His legs and body came to rest, hanging straight out, perpendicular to the post and parallel to the ground. So he could still do that! I knew this was how I would always remember Valera.

...

METHANE BUBBLES FLOAT but a second on the surface of a lake before they pop. After twelve years, my bubble of hope for a happy ever after with a Russian dancer had finally burst. But my love for Russia had not. Like a lake, Russia was a place I could count on as an escape from my home life, or at this stage in life, my lack of a home life. It was a place where I could become thoroughly enthralled by new discoveries of its people, its landscapes, and its moods. I decided I would delay the defense of my PhD dissertation and accept the offer to travel to the Lena River for another scientific expedition, alone.

On a late September day in 2005 I locked my cabin door on Chena Ridge and lugged my duffel bag full of winter field gear and boots down the forest path that led to my pickup truck. I took one last glimpse of my warm little cabin nestled on the hillside in a stand of golden-crowned birch trees. Lowbush cranberries in the understory had not frozen yet. I hoped I'd be back in time to collect and preserve some more for winter. Fireweed had already senesced. Its magenta flowers had bloomed to the tops of the stems, and now a few pink silks hung from open seedpods above a spiral of long bright orange, red, and green leaves. The poignant scent of Labrador tea lingered in my nostrils as I shut the truck door.

When I reached Yakutsk, seven-eighths of the way around the world from Fairbanks, I found the airport floor strewn with passenger freight, but there was no sign of my duffel bag. The bag contained much of what I would need for the expedition to the Arctic: winter field clothes, snow boots, and sampling supplies. For eight days I waited in Yakutsk for my bag to arrive. It did not. It was lost.

Dejected, I made my way along gray, dusty streets to see what I could buy at the Chinese market. Some of the coldest temperature records on Earth are from Yakutsk. The extreme cold makes the city's dirt roads very dry, despite their being underlain by permafrost. Clouds of dust floated around my feet as I walked. On the main road, murderous automobiles sent dust plumes so high above my head that I had to pull the collar of my shirt up over my mouth as a filter to give my lungs a break. At the market, I found cheap wool stockings, a sweater, hat and gloves, a coat, and a pair of boiled-wool boots. These would suit me well, but I would have to do without my bubble traps and sample bottles. At least shovels, brooms, and an ice spear could be borrowed, and I had my GPS unit, field notebook, and camera in my carry-on bag to map ice bubbles.

I felt disheartened. So far, Russia was not the uplifting experience I had been counting on. Garbage was strewn in the nooks of every street. Broken fences gave way to broken houses and broken lives. Everything looked worn out. Even the buses looked exhausted. I searched for something positive. "If anything is noble, if anything is lovely, if anything is worthy of praise or excellent . . . think upon these things." How did that passage from Philippians 4 go? I boarded a bus. That people survive generation after generation in this hole was noble, I supposed, but I didn't envy them. For a small measure of cheer, which was also admirable, they put plastic flowers in vases on windowsills of houses that were slumping into the moats of melting ground ice. I looked over at a solemn lady across from me. She too was clinging to the grimy rail above our heads as the bus tossed us to and fro down the rutty, dusty road. What would I see if I were able to step behind those big dark Yakutian eyes? What kind of iron were these people made of, I wondered, to survive here from day to day and year to year?

Beneath a heavy gray sky, I climbed a wooden plank onto the forty-foot cargo boat that would take me north along the river. This was the last trip any boat would make down the river this year. I was to sleep on a bench in the bow of the ship among stockpiles of cigarettes, Snickers, detergent, and staples the ship's crew would need to get them through the winter in the northernmost Lena River village of Zhigansk. I looked around my quarters for a place to set up an office for data analysis during the journey. The bench where I'd sleep did not have enough headspace for me to sit upright. I rearranged several fifty-pound bags of flour to make myself a seat. Cases of ketchup, canned fish, and peas would be my desk.

Winter was upon us when we reached Zhigansk four days later. Riverbanks were covered in thick snow. Ice was beginning to form along the river's edges. I knew that lakes froze up before rivers, so conditions should be perfect for my lake work. I watched the crew off-load supplies for the village and then went to the captain to ask when we could depart for the area north of Zhigansk, the area where I would be doing my research.

"The boat will not travel further. It will stay here," the captain answered in a gruff voice. "A storm is coming. We will all stay here now until the spring. Except you. In one week, an airplane will come to take you back to Yakutsk. Until then, you will live with the school director."

I was devastated. I had not come this far and delayed my dissertation defense not to collect any data. Was it possible to walk to some lakes from the village? The director said no, I must stay inside her apartment while she was at work. Outside was dangerous. Murders without cause were becoming more common in this village. Just that week a man had been killed when he went outside to empty his water pail. Someone had hit him on the head from behind.

The director's apartment was simple and warm, but I felt trapped. The Soviet-style cinder-block building stood on the bank of the Lena River. From the window I could see only about twenty meters of the river. The world was shrouded by blankets of falling snow. The wind blew bitterly. I thought about my cabin in Fairbanks. The wind didn't blow very often in Fairbanks. Sometimes leaving home helps a person put their life and dreams into perspective. Here I was, thousands of miles away and homesick for a home I'd not yet known. It wasn't my lonely little cabin I longed for. I loved my cabin, but what I longed for was a husband and a family. The idea of settling on a nice piece of land, having a little farm, and raising a house full of kids sounded so good to me. Desperate for fresh air, I ventured outside to stretch my legs, but I felt like a great big advertisement for crime walking around the streets even in the daylight, so I went back indoors.

I asked the director how I could get to some lakes. She couldn't fathom my intentions but said I could talk with her neighbor, an Evenki hunter. The hunter invited me into his kitchen to sit with him and his wife. He smoked into the woodstove as he told stories. Each November he was dropped off in the mountains by helicopter and didn't return home until April! I felt like I was looking at a superhero. This man was out in the wilds with his tent and snow machine (the Alaskan term for "snowmobile") all winter long, often alone or with his son. It didn't matter that for several months the sun never rose above the horizon and Siberian temperatures plummeted to minus 70°C! His tent was made of heavy canvas and it didn't have poles. He took a few minutes everywhere he went to chisel new tree trunks for poles. He killed every animal he came across. In a single winter he harvested more than a hundred sables and forty-five moose. He killed mamas and babies alike. I said I thought this was sad.

"We are not a sensitive people," the hunter answered, tapping his cigarette into the stove.

The hunter's skin looked like burlap, a strong contrast to his wife's skin, which was more akin to the mare's milk on the table—soft and creamy. She stayed home all winter, I explained to myself, while he was a ghost husband. I thought about all those harsh winters in the wilderness.

"It is incredible that you haven't died out there," I said.

"Each person has a certain time and way that they are supposed to die," the hunter answered. "My time has not yet come."

Then he turned and looked intently at me. "Today they found the body of one of three young men who recently drowned in the Lena River. They were ecologists from this town, probably drinking—ages twenty-five to thirty years old. It was terribly sad, especially for the man's mother, who was fifty-four years old and is now left alone in the world. She had two grown sons. Both died this summer. The first one was beaten to death here in the village, and now, in the same summer, the second one died by the river." The hunter tossed the butt of his cigarette into the stove and added, "Not even war is so cruel to a mother."

I paused to feel the weight of his sentiment and then broke the silence. "I came here because I want to study lakes," I explained. "Is there anyone who would be willing to take me out to some frozen lakes? I will pay them."

The hunter thought for a few minutes and then told me that there were some lakes thirty kilometers from there. A father and son, by the name of Titov, had an izba (hut) near those lakes. The native Evenki people of this region were nomadic for thousands of

First snow

years, until this most recent generation. This particular family used to raise cows, pigs, white foxes, and chickens out there. Since the breakup of the Soviet Union, the farm had fizzled away due to a lack of government support, but the father and son still went there frequently as a base for hunting and fishing. He thought they'd be willing to take me out for a sum of money.

Later that evening I was called to the school director's front door. There stood her neighbor, the Evenki hunter. "The Titovs will take you to their izba tomorrow after lunch. Be careful," he added. "This territory is a graveyard."

A bit frightened and not knowing what I had gotten myself into, I packed my field gear, hid my valuables and computer in the baby's room at the director's house, asked to borrow a broom and shovel for fieldwork, and then scurried around town to buy canned vegetables, pasta, chocolate, garlic, and other essentials for several days out in the woods. Then I waited. At 2 p.m. the hunter arrived and loaded his snow machine with my gear. I rode on a sleigh on top of a canvas full of all my borrowed gear and food.

The Titovs lived on the edge of town. When we arrived, their snow machine was strewn apart as the father and son tossed different oily tools around in their dark garage, making their final repairs before our journey. I was skeptical that their machine would make it; they were a very poor family, but they knew how to fix things, of that I was sure.

The Titovs led the way several hours by snow machine through a forest of larch and birch. It felt good to get out of the village, away from civilization, away from danger, into the comfort of the wilderness. I breathed freely for the first time, almost unaware of the plumes of unfiltered exhaust blowing into my lungs behind the snow machine.

The Titovs' farm was quaint but decrepit. The deserted pig barn made of mud and moss was now covered in snow. We stepped into the darkness of their wooden cabin. As my eyes adjusted to the dim light, Simeon, the father, and Vasya, the son, rushed about clearing away dirty dishes. They stashed a pot of leftover frozen soup, swept the floor, and moved the furniture in such a way that I would have a bit of privacy on one side of the room, all of which I thought was very considerate.

Complete darkness fell. I rummaged through my backpack and found the little LED headlamp I'd purchased in Fairbanks just before this expedition. But once the Titovs had their kerosene lantern burning, I left my headlamp on my bed and started cutting onions for supper. I turned around a few minutes later when Simeon spoke to me and, to my surprise, found him walking around the cabin in *my* headlamp. I thought it was rather strange, and possibly a bit rude, for this man to take my headlamp, but I didn't say anything. After a while, he started clicking the switches between light levels and asking me if mine had different efficiencies. I said yes, just as I realized it was *not* my headlamp that he was wearing. With all the dirty rags, animal skins, naked metal mattress springs, and Soviet apparatuses in the cabin, the last thing I had ever expected was that Simeon would own a headlamp just like mine! But he did. I told him I thought he was wearing mine, and we both erupted in roaring laughter that broke the ice.

In the middle of the night, I was startled awake by the sound of different laughter, evil laughter. My eyes had shot open at the first sound of cackling, but in the pitch-black of the cabin, I could see nothing. I lay frozen stiff, petrified in my bed. I didn't dare reach for my headlamp. It seemed best to remain dead still and let my ears do the discerning. All was quiet for a while. Then I heard something take a deep raspy breath before letting out another long trail of high-pitched laughs. *Heeeeeeeeeee. Heee. Hee. He. Heeeeee.* The cackling came again, and again. I wondered if whatever was laughing could see me. Could it tell that I was scared? What was the purpose of this base laughter? I'd heard stories about evil spirits terrorizing native hunters in the Cherskii region, nearly a thousand miles east of where I was. To ward them off, the natives would feed horsehairs to the fire. Such stories had always seemed like old Yakutian wives' tales to me. Now, paralyzed by fright in the pitch black in the middle of nowhere on Earth, I was determined not to let terror get the better of me. "For God has not given us a spirit of fear, but of power and of love and of a sound mind" (2 Tim. 1:7, NKJV). More likely, I figured, the laughter was coming from one of my hut companions. I heard footsteps shuffle across the floor. The door opened and the laughter went outside. That night didn't end fast enough for me, but somehow I eventually fell back asleep.

In the dim light of dawn, I awoke to more shuffling footsteps on the cold, dirty wood floor and the creak of the woodstove latch. Someone was adding wood to the fire to stave off the icy cold seeping in through the cracks in the cabin's wall. I quickly put my field clothes on and walked around the small divider Simeon and Vasya had made for me. They both were already up and getting ready for the day. We greeted one another, and I searched their faces for any sign of having heard the evil bouts of laughter in the night. I didn't dare ask them about it, but my question was answered soon enough. I spun around when I heard another raspy draw of breath and the beginning of the familiar sinister laugh.

What I saw wasn't laughter at all. It was Simeon. He was coughing, desperately wheezing and hacking. He pulled a cigarette out of his pocket, lit it up, and went outside to smoke, stifling the evil cough. My heart went out to him. It sounded like it deeply hurt. I looked up at Vasya. He was busy making breakfast and didn't say anything. When we started to eat, they explained to me that Simeon was dying of cancer. He was also suffering from both tuberculosis and diabetes. I thought Simeon did look old and rather weak, but they said he was only in his late forties.

For the next three days, there was never a dull moment. As long as there was daylight, this father-and-son pair were busy fixing or preparing something. After nightfall, they sat in lantern light sipping soup and tea and laughing with me as we learned about each other's ways of life.

"Growing up in the Soviet Union," Simeon admitted, "I always believed that all foreigners were spies. I never imagined an American girl would someday come to this homestead."

I smiled at him yet couldn't help but feel that he still had not entirely let go of his doubts about me.

The Titovs were eager helpers with my fieldwork on the lakes too. Their assistance enabled me to observe important and striking differences in bubbling in the different regions of Alaska and Siberia. I began to realize regional differences in the supply of permafrost soil carbon fueling the methane production caused different amounts of bubbling. There in Zhiganks, the soils were sandy and contained little carbon. As a result I saw far fewer methane bubbles than I had previously in the carbon-rich yedoma permafrost regions near Cherskii and Fairbanks.

Vasya shoveled snow along 3-foot-wide by 150-foot-long transects. The snow was only half a foot deep, but it was sopping wet, making the shoveling grueling work. It didn't help that the shovel was made from a sheet of bent metal wired to a crooked tree-branch handle. But Vasya shoveled skillfully, as efficiently as he fixed the snow machine and walked straight lines across lakes to fish with his dad. There was not one wasted movement in his body. He was graceful and strong, soft and handsome. Whenever I said "*Spasibo* (Thank you)," he answered a quiet and sweet "*Pozhalusta* (You're welcome)."

My romantic thoughts quickly dissipated as, standing at the head of one of the snow-shoveled transects, Vasya told me about his wife. "She is from another village. We married three months after meeting. But that was six years ago now. Our eldest daughter lives with us, under the same roof as my parents. Our younger daughter lives in a remote village with her maternal grandmother."

"She does?" I gasped, not fathoming that parents would let a little child go.

"Yes," Vasya explained. "Simeon's house is already full of grand-children since some of my sisters had children they cannot afford to house. So, the burden of our younger daughter was spread to the other grandparent." Wow, their lives were sure different from ours in the US, I thought to myself. I wondered what kind of a toll these separations had on their hearts.

While I classified and mapped methane bubbles, the father-and-son pair set fishing nets beneath the ice. With bare hands in the wind and snow, they pulled tens of round, white fish up in their nets and laid them out on the surface of the lake to freeze. Their harmonious labor as a father-son duo struck a sensitive chord in my heart. Words were few as they worked. They knew what to do and didn't need words. I wondered what it would be like to know my own father so well. These two did everything together, but since his father had fallen sick, Vasya took most of the burden, yet in a way that did not diminish his display of respect for his father. I wondered what it would be like for Vasya when his father died. To live and work so closely together almost seemed more intimate than a husband and wife. Surely these two must have a tight bond of love. I had never come across anything like it before, and I wondered if it was a rare thing.

Simeon and Vasya's work was frequently interrupted by convulsions of coughing at the fishing hole. Simeon's cough sounded painful. He smoked a few packs of cigarettes a day and often relieved his cough by lighting up. During Simeon's bouts of pain, or during his regular shots of insulin, Vasya kept silent at his business, though I imagined his heart must be breaking. How could Simeon himself be so loving and fun when he was dying so painfully from the inside? It was a mystery to me at that time. But later, as a parent myself, I would realize that when we know our time with loved ones is limited, then compassion, laughter, and joy become the most important interactions.

Since they had shared with me that Simeon was dying and since I was curious about what kind of faith might lie behind the love in their relationship, I decided to ask a personal question, a question I was still working on within myself.

"Simeon, what do you think about God?" I asked during our last night in the cabin.

The thin, weathered man thought for a moment and then answered, "It is the division between all different kinds of people who call themselves believers that is a barrier to me," he said. "How could people be so divided if there is one true God?"

"I'm troubled by that same question," I admitted. "But what if God is larger than the weaknesses and limited understanding of men?"

We never finished that conversation, but I kept wondering about the impact society had on a person's beliefs. Before traveling to the Titov homestead, I had met Simeon's father, a frail 112-year-old man who sat on a drooping mattress in a high-rise apartment in Zhigansk, his face lit up by the warm glow of afternoon sun streaming through the window. Great-grandfather Titov's room was bare except for an icon of Christ on the cross behind his bed. He was too blind and deaf to converse with me, but I couldn't help wondering if he was a believer from pre-Soviet times. Now, having become acquainted with his posterity, I thought of him again. Simeon was a biblical name. Perhaps Simeon's father had faith in God, but Simeon's own unbelief was the product of Soviet atheist doctrine, infused in him like his distrust of foreign spies. How much of my own unbelief, I wondered, had been fueled by a society that applauded human achievement and laughed at obsolete notions of God?

I had not known what to expect when I agreed to go with these strangers for three days out to their homestead. They could've been uncomely people. What a surprise I'd had in store! They were kindred spirits. They had been as scared of me as I was of them, but we had stepped out in faith and overcome our fears. We had discovered joy in one another's company despite our differences. The Titovs were the first people in my life who had showed me what true family love could look like. I saw it again when I spent my final day with their family, cooking pizza for them, their wives, and a couch full of uncharacteristically quiet children. It was clear that a person did not need to be materialistically rich to have a wealth of love in family. In fact, I wondered, do material riches get in other people's way? The Titovs laid themselves down to help one another. They put each other first. Wasn't this the definition of love? Looking past the leaky tin roof of their house and the living room couches that served as beds for three generations at night, it seemed to me they were mighty rich (see center insert photo 10).

I had mixed feelings about my return to Fairbanks. On one hand, I looked forward to feeling safer in town, sleeping in my own bed, lighting the woodstove of my cozy cabin, playing my cello, enjoying my friends, and skiing on the forest trails outside. But on the other hand, I knew that the next months would bring big changes as I defended my dissertation and prepared to take my first job as Dr. Katey Walter. When I walked back up the path to my cabin, after driving myself home from the Fairbanks Airport, I found a welcoming party. A mother moose with three babies was browsing on the now-frozen sweet peas in the round wooden planters on my deck. I thought of the Evenki hunter getting ready to start his winter work in Zhigansk. I missed the Evenki already, but I was glad my visiting moose were here and safe from harm.

...

> Some wandered in desert wastes, finding no way to a city
> to dwell in; hungry and thirsty, their soul fainted within
> them. Then they cried to the Lord in their trouble, and
> he delivered them from their distress. He led them by a
> straight way till they reached a city to dwell in. Let them
> thank the Lord for his steadfast love, for his wondrous
> works to the children of man! For he satisfies the longing
> soul, and the hungry soul he fills with good things.
>
> Ps. 107:4–9, ESV

In Fairbanks in December the sun rises above the southern horizon around ten thirty in the morning and sets a little after two thirty in the afternoon, giving four hours of meager daylight. It is not a light that warms you, and yet it is a light. Despite a temperature of 35 degrees below zero, I yearned to feel daylight on my skin. I took off the black, face-warming mask and let the sun's weak rays hit my cheeks. Immediately my breath started to freeze on my eyelashes and the strands of hair that had escaped from my braid around my face. I had already worked up a sweat on my cross-country skis. I knew I should not linger unmoving for too long or the cold would sink into a place I could not shake it. I was miles away from home and my wristwatch read 2:08 p.m. I'd have just enough time to ski back through the forest to the clearing by the Tanana River to watch the sun go down and then cover the final mile or so climbing the hill to my cabin in the sweet yellow-and-pink light of Arctic alpenglow before the next twenty hours of silent darkness would set in again.

Winter sun in Alaska

Reaching my cabin in the afternoon twilight, I propped my skis upright in a pile of snow by the door. One trick I'd learned to staying warm in Alaskan winters was to bathe immediately after running or skiing. Letting evaporative sweat linger was an invitation to an inner chill. I had no plans to drive ten miles to town, where showers were freely available in the basement of the Arctic Health Research Building. I would have to bathe at my cabin instead. This meant taking the pot of hot water off my woodstove, mixing it with snow to cool it down, and pouring the warm water over myself outside on the edge of the deck, where the water could run through the wooden slats to the ground below. Although it was not completely dark yet, my cabin was well enough hidden from the road that I did not hesitate to bathe right out front in the nude. I had wetted and soaped myself and was just starting to pour the warm water rinse over my shoulders when vehicle headlights cast their light on me from the one bend in the neighbor's driveway where a person could possibly see what I was doing. I quickly dumped the water and turned to dash back into my cabin. Suddenly I slipped and sprawled out naked on the icy wooden slats. The vehicle moved on, and as embarrassed as I was, more for the slipping and sprawling than for having been caught in the nude, I went back inside and curled up on my towel by the woodstove, letting the radiant heat warm my skin.

The lights were off in my cabin. An orange glow emanated from the woodstove's glass, but the brightest light was that of the moon reflecting off birch tree trunks outside. I leaned forward and looked up through the window glass at the illuminated sphere of the full moon hanging still above the denuded crown of birch forest. Loneliness again pulled heavy on my heart. I was a human, and how I longed to be loved by somebody. The sting of lifelong disappointment in my relationship with my dad throbbed in my heart. Love is sacrifice. Why hadn't he sacrificed certain things along the way for me, his child? Why hadn't he provided for me at times when I'd asked him and thought I'd needed him most? It seemed his new marriage had sealed me only further outside his heart and home. Then there was Valera. My naïve hope of a life with him had been dashed, and a long list of other failed romantic relationships trailed into my past. Here I sat, empty and alone. In my despair, I uttered a prayer: "God, is there not one person who will love me?"

Then, in an inner voice, God answered me: *"Katey, there is a man who loves you so much that he's willing to give up everything for you. Everything. And he's already done it."* I knew instantly of whom God was speaking. It was Jesus. During endless Sundays in church, I'd heard about Jesus dying on the cross to forgive my sins. But he had died for everybody in the world, not just for me. I was just one of billions of other people. I wanted someone who would love me specifically. I wanted to be special to someone. God spoke to me again: *"Katey, do you know that I love you? Do you know that even if you were the only one, I still would have sent my son to die for you? Will you accept that you are worthy of that?"*

Would I accept that I was worthy of that? I realized that when God spoke, I didn't really have a choice. Of course I was going to accept this. He himself was telling me he loved me. He was freeing me of the expectations I had laid on my father and on any human relationship. I understood now, in a way I never had before, that God loved me perfectly, like no human ever possibly could. This was enough. In those moments, God filled a gaping hole in my heart. I was healed from the pain of my past, and in its place was peace. At that moment, I started to love my Lord. Not in my head but from within my heart. I recalled a Bible verse I'd read before: "We love him because he first loved us" (1 John 4:19, NKJV). For the first time in my life, I could say this was true. But I, self-centered Katey, still had a lot to learn, especially about what it meant to love God and people. True love was much more than a feeling. "Greater love has no one than this, than to lay down one's life for his friends" (John 15:13, NIV).

...

SEVERAL MONTHS LATER, in March 2006, I found myself staring at the last remaining leaves of my ficus tree as they detached from their branches and fluttered to the linoleum floor of the cottage I was renting in Cordova, Alaska. After finishing my doctoral program, I had taken my first job in this Southcentral commercial fishing town. During the seven-hour drive south from Fairbanks to Valdez, and then the thirteen-hour Alaska Marine Highway ferry ride across Prince William Sound from Valdez to Cordova, the limbs of this unfortunate tree had been caught in my truck's door and exposed to subfreezing wintry temperatures and windchill. As my tree dropped its final leaves, I was inclined to throw it away altogether. But I called my former Fairbanks cabin landlord, the best plant doctor I knew. "Wait," she said. "Give the tree time. It will come back."

I sat at my dining room table and looked out the window toward Cordova's boat harbor. The harbor's still water reflected the gray-blue sky, except where sea otters flipped, churning the water's surface when they dove to forage for shellfish at low tide. Out on the farthest pier stood the Prince William Sound Science Center, my new workplace. The attractive wooden building had windows on all sides and a long bright-blue roof that sloped toward the water. In a second-story alcove with a window looking out on the boat harbor, town, and mountains, was my office. In a few minutes, after answering another university-related email from my kitchen table, I'd be heading that way by bicycle to work.

I was *Dr.* Katey Walter now. For the first time in my life, I earned four-digit paychecks, enough money that I could start to think about buying my own house. This new title of doctor had helped me land my current job as research program manager of the Prince William Sound Oil Spill Recovery Institute (OSRI), which was administered through the science center. My responsibility was to oversee the budget of $1.2 million mandated by Congress in response to the 1989 *Exxon Valdez* oil spill. OSRI's charge was to identify and develop techniques for dealing with oil spills in the Arctic, and to promote research, education, and outreach that would improve understanding of the long-term effects of oil spills on the environment and the people who depend on marine resources.

I loved learning about coastal ecosystems and establishing productive relationships with OSRI's board of directors as well as different sectors of industry and state and federal agencies. Such breadth of interaction certainly had not been part of my doctoral research tucked away in the Siberian Arctic. I also loved the opportunities this new job offered for travel to various coastal communities of Alaska, to Washington, DC, for lobbying, and to Norway, where the most cutting-edge research on oil spills in ice-covered waters was being conducted. And yet, as I delved deeper into my new area of work, the waves of my former Siberian methane research had not yet dampened into ripples. If anything, the waves of opportunity were mounting higher. There were tall stacks of my field notebooks, filled with data I still felt compelled to turn into manuscripts. The journal *Nature* wanted new figures for finalizing my first publication. My PhD dissertation had just won the first-place award of the US Council of Graduate Schools in the field of mathematics, physical sciences, and engineering. Colleagues, among whom I could now rank as a fellow "doctor," offered me opportunities to participate and even lead new research proposals to the NSF and NASA. And just that morning I'd received an email inviting me on an all-expenses-paid trip to England to defend my ideas about permafrost thaw and methane to the Royal Society of London. Alongside the private defense, they asked that I prepare a new manuscript to publish in their journal, *Philosophical Transactions of the Royal Society A*. I knew already that putting a new manuscript together in a matter of only a few months would be a steep task, especially with my new full-time job at OSRI, and yet I had never turned down an intimidating and constructive challenge. Besides, this would be a perfect opportunity to summarize the data I'd collected on lakes in Zhigansk and northern Alaska to compare to my original Siberian lake data set.

I had long wanted to know how representative Siberian lakes were of other lakes in the Arctic, and now I had data from seventeen lakes representing not only Siberian thermokarst but also various glacial, alluvial-floodplain, and peatland lakes. My mind raced about this new opportunity as I shut my laptop and hopped on my bike to coast down to the harbor to work.

Fulfilling the responsibilities of a full-time job at OSRI during the day and pouring over my methane research at night meant that I was getting very little sleep. But sleep didn't seem to matter. Adrenaline pumped by ambition and opportunity pulsed through my body. The world was at my fingertips. There was so much to see, so much to explore, so many challenges to undertake. But the daily transition I needed to make between oil-spill work and methane research required some kind of bridge.

I had discovered that mountain trail running was just the bridge. It drained my mind of all thoughts other than where to place each foot along rugged paths that led up into the mountains from the fjord in which Cordova was nestled. I remember one run in particular: after work I peeled off my business clothes, leaving them in a temporary heap on my bedroom floor, and dressed as quickly as I could in my sports bra and sweats. I didn't mind that it was raining. Snow had turned to rain only a few weeks prior, and the rain had waxed and waned now for days. Sometimes gusts of wind would lift and throw the waterdrops back up into the sky so that it looked as if it were raining upside down. But that afternoon, the rain was only a drizzle. The small town was socked in by thick gray clouds. I'd been in Cordova long enough to know that if a person waited for the sun to come out before venturing outside, they'd hardly ever go. The thing to do was don my rain gear and get out there.

My calves burned as I forced one foot and then the other up the steep, craggy incline of Mount Eyak, the tallest mountain I could easily access from my front door. With each step, water pooled between my old gray tennis shoes and the soggy mosses that gave way to brown soil, exposed cedar roots, and rocks on the narrow footpath. *Don't even think of stopping,* I said to myself as I ran upward, one foot slipping on a sharp, muddy rock in the trail. I put my hand out to catch myself, just missing the stinging spines of a massive devil's club shrub.

This day I was in for a surprise. No sooner had I wiped away the sweaty stream of water that had stung my eyes than I felt a brightness of sunlight hit my eyelids. Strands of fog dissipated, and strong beams bombarded my face. Not far from the mountain peak, I stopped. My heart pounded. I stretched out my arms and threw back my head. The sun had never felt so good before in all my life.

I couldn't believe that here, at the top of Mount Eyak, patches of moss and grass were actually dry and basking in the sunlight. I closed my eyes. I was in no hurry to escape this sensation of ecstasy. To my mind came the woman pictured on the Celestial Seasonings tea box at home. She was standing on top of a mountain with her white dress and hair blowing in the wind. It seemed as if I could hear an angelic choir singing, but when I opened my eyes and looked around, I realized the only sound I could hear was the bestial roaring of Stellar's sea lions resting on buoys in Orca Inlet 1,200 feet below me.

My mind turned to a woman I'd read about in the book of Isaiah over my cup of morning coffee. She was referred to as Babylon and called a "Lady of Kingdoms." She said, "I shall be a lady forever." She was given to pleasures and sorceries and an abundance of enchantments. She dwelt securely in her place in the world and said, "I am, and there is no one else besides me" (Is. 47:5–8, NKJV).

It struck me that I was becoming like Babylon. I was basing my sense of security in my education, competition, success, and career advancement. Isaiah goes on to say that none of these things, nor the people with whom Babylon had labored from her youth, would be there to save her. Disaster would come upon her, and she would not know how to conjure it away. My mom had talked about this before. "Katey," she had said to me one day on the phone, "where will those rungs of your career ladder be when the storms of life hit? They will not save you. The only one who can uphold you is the Lord."

Above the clouds

I thought of Isaiah 48. "Hear this," the word of God continued. "I have not spoken in secret from the beginning; From the time that it was, I *was* there . . . I am the Lord your God, who teaches you to profit, who leads you by the way you should go. Oh, that you had heeded my commandments! Then your peace would have been like a river" (Is. 48:12–18, NKJV).

I looked around at the sea of clouds below me. I was aware that my head was becoming puffed up with pride in my scientific success. My achievements were mounting to earthly greatness and fame. Taking pride in this was wrong. And dangerous. As the saying goes, pride goes before a fall. I needed to seek God in all things. He would make my path straight.

The clouds were a palette of rose, yellow, and white through which jutted an archipelago of mountain peaks. The shoulders of the tallest peaks were draped with glaciers, marking their proximity to the vast ice fields of the Chugach Mountains beyond. I wanted to stay up there in that bright, colorful wonderland where, once again, God was softening my heart and turning it back toward himself. But there was work to do below, only now I knew I must seek to be more humble.

On strength not my own, I went to London and defended my ideas about the role of methane bubbling in various types of Arctic lakes. I showed how bubbling was more prevalent than we had thought, and that taking it into account made methane emissions from northern lakes globally important.[1] Considering the magnitude of carbon locked away in permafrost, my calculations also suggested that complete thaw of Siberian yedoma permafrost alone could release 49 billion tons of methane (today's atmosphere contains only about 5 billion tons of methane). Additional methane would be released from other non-yedoma permafrost soils outside Siberia, but the magnitude of emissions at any one time, and therefore the impact on Earth's climate, would be a function of the release rate. I acknowledged that there were huge uncertainties in the rate of thaw and release. Nonetheless, my calculations suggested that, in the future, a large pulse of methane was expected to come from permafrost thaw. Unfortunately, I had used the term "methane time bomb" in my paper. This was an overstatement of the problem that was subsequently repeated by many media streams around the world. My calculations also suggested that in a far, far distant future, when permafrost no longer dominated the Arctic, the number of lakes, and therefore their methane release, would be much lower. The timescale of this trajectory was uncertain, but this paper helped to set the framework for a strong and growing body of research that today aims to better refine the permafrost carbon feedback to global climate warming.[2]

By May 2007 it was apparent that my calling was in academia.

A person can go on only so long burning a candle at both ends without even getting paid for it to conduct research in an area they are passionate about. When the University of Alaska Fairbanks offered me an International Polar Year Postdoctoral Fellowship and then an irrefusable start-up package as an assistant professor of research to continue my studies of Arctic methane, I could not turn it down. My heart lamented the thought of leaving the ocean and fjords, the smell of salt water, the sound of seagulls, and the sights of eagles, otters, sea lions, whales, and fishermen, but I was beginning to swallow the undesirable truth that a person cannot have all things at all times. Besides, if I hadn't accepted this opportunity to move back to Fairbanks, I can't imagine how I ever would have met Peter.

ELEVEN

Peat Cakes and Wedding Cakes

Come on, Katey, of course you'd be welcome. The reception is at 6 p.m. at Mushers Hall. Everybody will be glad to see you."

The man whose Chena Ridge house I was in the process of buying during a brief trip to Fairbanks, before moving my belongings up from Cordova, was trying to convince me to crash the wedding party of Bobby Bursiel and Morgan Peterson, peers from my earlier life as a graduate student in Fairbanks. The man was the Bursiels' neighbor and friend, and he was catering the wedding.

"Oh, all right," I said. I'd been to only a handful of weddings in my life. I knew little about wedding etiquette, but something told me it was not okay to go to a reception to which you have not been invited. On the other hand, this was Alaska, a place where freedom reigned and rules often didn't apply. I had plans that afternoon for sand volleyball, but the wedding reception wasn't until evening. Surely there would be lots of people at the wedding I'd know from previously living in Fairbanks, I told myself, pondering the opportunity. This would be a great way to reconnect, I decided. Besides, the caterer had sort of invited me, hadn't he?

Just before 6 p.m. I pulled over at a gas station to wash the sand

and sweat off my body with paper towels and change into the nicest clothes I had along on my trip: a pair of tight-fitting brown cotton culottes and a navy-blue blouse. I let my ponytail out of its band and loosened my curls down my back with my fingers. This would look all right with some jewelry and lipstick, I thought. I jumped back into my black two-seater Toyota pickup truck and drove to Mushers Hall.

Vehicles filled the large gravel parking lot outside the hall, a log building set on forty-five acres of green grassy fields with thick birch forest and mushing trails surrounding it. With each step forward I wanted to turn around, dash for my truck, and drive away. But I forged onward, beating down my self-doubt. I felt like an imposter walking through the front door. Immediately faces I recognized, from the wedding party itself, brightened up when they saw me. This emboldened my steps. I walked straight over to say hello to Bobby and his family. If this was to be a party where people simply stood and mingled, then my presence might not cause too much of a ripple, I assured myself. I started to feel a little more at ease.

Behind Bobby, a blond young man absorbed in another conversation caught my eye. The man looked athletic, with broad shoulders, and he stood a good head taller than me. His neatly pressed tan suit and clean-shaven face gave him away immediately as a non-Alaskan, and so, I told myself, I should not be interested in him. Alaska was the only place I wanted to be. I'd just signed a contract for my dream job as a research professor at the university, and I was buying a house. I was enthralled with the rungs on my career ladder, so what was I doing glancing back again at this man with a handsome face and features? I hadn't come here with any hopes or expectations of meeting someone new that evening, but now I found myself attentive to the question of whether or not this curious young man had a date. I didn't see any sign that he did.

The call was made for people to go outside and get some food from the potluck-style row of tables piled high with homemade Alaskan dishes. Again, I relapsed into a feeling of awkwardness for having arrived without an official invitation. I lingered to get my food last. Finally, with a plate full of grilled salmon and halibut and fresh Alaskan greens and berries, I looked around for a place to sit. Every seat under the tents appeared taken. Of course they were. Surely the hosts had carefully calculated seating for their invited guests, and I was not one of them. Just as I started to feel even more humiliated, wondering if I should go into the kitchen and eat my food next to the caterer, I noticed an empty place in the far south-east corner of the tents. And of all things, it was a seat right next to the blond man in the tan-colored suit.

I ambled over, shy, embarrassed, excited. "Is this place available?" I asked.

The man looked up at me with pleasant blue eyes and smiled courteously. "Yes, it is. Please, feel free to take it."

"Thank you," I said, with poise intended to hide my immense relief at sitting down and blending back into the crowd.

We regarded each other with anticipation.

"My name is Peter," the man said presently in a friendly sort of way that was not overly gregarious.

"I'm Katey," I answered. The invitation to converse with this intriguing and unfamiliar man seemed to absolve me of all feelings of guilt and impostordom. I was completely at ease with myself again, confident and cheerful.

"How is it that you know Bobby and Morgan?" I inquired.

Peter paused. His brow furrowed while he focused his thoughts. Then he turned his eyes to mine and explained with a relaxed expression, "Bobby and I were roommates at St. Olaf College in Minnesota. He often came to my family's farm when he was a student. Our farm is only an hour away from St. Olaf, so it was easier for Bobby to come there for some holidays than to Alaska. We shared a love of the same books and baroque music, so naturally we became good friends. Bobby asked me to be the best man in his wedding."

Oh boy, I thought. My guts felt as though they'd fallen into my lap. He was the best man in the wedding, and I wasn't even invited. My imposter feelings rapped at the door, but I kept them out and forced my mind to move on.

"That's right. Bobby plays the cello, doesn't he?" I recalled that Bobby's family was musical and that his bride, whom I knew much better than Bobby, was both a fiddler and a scientist. I didn't wait for Peter to answer. "Is this your first time in Alaska?"

"Yes, Bobby is a very good cellist." Peter answered my questions in order. "It's my first time coming here in June. Usually, I've come in August. Ever since I met Bobby, which was . . . let me see . . . ten years ago, I've spent many Augusts here helping him and his family at their air service."

"Do you fly planes?" I asked, increasingly intrigued. I half expected him to say yes. Alaska had the highest per capita ratio of pilots in the nation since planes were often the only way to get to remote places in a state characterized by wilderness and very few roads.

"No. I've always wanted to get my pilot license, but I just help Bobbly load cargo, food, and supplies into planes bound for rural villages. We always set aside some time to canoe various wild rivers in Alaska too. He and I both love paddling."

This interest in adventure added excitement to the admiration I already felt for his taste in music and reading and his strong, work-worn hands. My eyes followed the chiseled lines of his face and I thought that it wouldn't be so bad to look at those long cheeks and that strong jaw for a lifetime. But then I caught myself. I must rein in my interest if he didn't answer a certain question the right way. I didn't bother trying to think about how to phrase it. I just blurted it out.

"Bobby's family goes to church, I recall. Are you a Christian?"

At thirty-one years old and starting the next serious step in my career, I was no longer interested in entertaining romantic relationships that had no possibility of leading to marriage. I'd broken up over faith disconnects before. Making sure we shared the same faith seemed a more important hurdle to leap than the inconvenient fact that we lived in two different states.

"Yes, I'm a Christian," Peter said.

"Oh, I am too," I added in a nonchalant voice, hiding the intrepid excitement I felt for the opportunity opening before me.

Hoping he didn't mind the highly personal and perhaps out-of-place question I'd just posed to him, I moved the conversation on.

"So what do you do in Minnesota?" I asked. Neither of us had made a dent in the food on our plates, so I started eating.

Peter left his sandwich untouched on his plate and explained that he helped on his parents' farm. He was an only child, and they would be turning seventy this August, so they needed his help.

"What kind of farm do you have?" I asked, realizing I didn't know enough about farming to even know where to start asking an intelligent question.

"We spend most of our time raising crops, about three thousand acres of corn and soybeans. We plant small amounts of peas, sweet corn, and wheat. We also raise pigs and cattle. By August, the crops are so tall that there is not a lot we can do but let them go to full maturity on their own. That is why it is a good time to come up here to Alaska." Reaching for the first time for his sandwich, Peter asked, "What is it you do?"

"I'm a professor," I said, proud of my new position. "After graduate school, I worked as a program manager at the Oil Spill Recovery Institute in a little fishing town called Cordova. I lived there the past two years, but now I'm transitioning back to Fairbanks to start a job as a research professor at UAF. I study methane in lakes."

"Oh, so you know about methane?" Peter asked with a gleam in his eye. "What in particular do you study?"

I told him about permafrost thaw and lake formation and about how methane bubbles composed of ancient-permafrost carbon were escaping into the atmosphere. "It seems you know something about methane too," I remarked. "Did you also study it in school?"

"No." Peter laughed. "I studied engineering but got my degree in history. I'm interested in methane because we raise pigs that produce a lot of manure and methane. I've been trying to find out if there is a way we can harness the methane for energy."

"Did you go to graduate school?" I asked, curious about his level of education. He was certainly thoughtful and very well spoken, characteristics I thought might reflect academic ambition. It didn't really matter to me if his academic level matched mine or not. Having spent nearly all my life in school surrounded by other students and teachers, I enjoyed the novelty of people from the "real world." And I knew accolades didn't necessarily reflect intelligence. After all, Sergey was the leading expert in permafrost carbon science with multiple publications in *Nature* and *Science*, but he had no graduate degree. I was curious to know more about this man next to me.

"I did take some graduate courses in Renaissance history at Oxford University in England," Peter said. "But I didn't pursue a degree. I studied out of interest more than anything. I had already decided by the time I finished St. Olaf that farming was the direction I wanted to pursue. Oxford was an opportunity to become more well-rounded before committing my efforts solely to farming. I really enjoy farming. I get to work with my hands, spend a lot of time outdoors, and I like the diversity of responsibilities and challenges in farming. It involves everything from soil management and plant physiology to mechanics and economics."

It was not evident to me how our commitments to vocations in two different places and fields of work would ever possibly be compatible, but I'd learned long ago that I didn't need to have all the answers at once. Following my instincts and walking in faith had proved well so far.

Before the evening ended, I asked Peter if he'd like to extend his stay in Alaska to join me in Cordova for a few days. After that we could put my truck full of my belongings on the ferry and drive it back up to Fairbanks together. I knew this invitation would coincide with my dad and Sandy's upcoming visit. I hadn't seen them in several years. Our relationship was cordial, but not close. To them I was still an emotionally irresponsible and self-centered girl. I'd gone against my dad's advice to pursue a future in academia. Their impending visit would be a chance to show them my life in Alaska and that I had grown up. It was an opportunity for them to get to know a little more about me in the place I loved. It was just too bad their visit would overlap with my invitation to have Peter join me in Cordova. I wasn't so sure they'd be happy to have a stranger present on their Alaska trip, but it would be for only a couple of days, I told myself.

Not knowing he was impinging on their trip, Peter agreed to the plan.

The sun shone for six glorious days in Cordova. True, Dad and Sandy were disheartened to learn I'd invited a stranger to join us, but I figured they would get over it, and besides, me finding a husband seemed more important than a minor short-term offense to them. Dad, Sandy, Peter, and I ate a picnic across the river from Childs Glacier, watching huge sections of the ice wall calve. Chunks of ice splashed down into the river and sent sprays of water all the way to where our feet were outstretched on the rocky beach among the magenta fireweed. As my romantic attraction to Peter heightened,

I was glad to see that my dad seemed interested in Peter's stories of farm life. After Dad and Sandy left, Peter and I hiked the mountains alone. With a tent and food on our backs, we climbed much higher up Heney Ridge than I'd ever gone before. We glissaded down patches of snow that had not yet melted, landed in fields of June wildflowers, and bathed in the ice-cold meltwater lakes. All along the way Peter asked questions about my family and my past. His questions were wise and insightful. By them I discerned that he was a very thoughtful man and a man with sound judgment. I could tell he was more careful than I had ever been about whom he would get romantically involved with.

My admiration for Peter grew as I learned more about him. Raised as a single child of older, well-educated parents on a large, rural farm, Peter had spent most of his life in books when he was not working on farm projects. His knowledge and skills were unlike anything I'd ever encountered in a person my age. "The *Franz Ferdinand*," he explained to me, looking at the name painted on the stern of an old wooden fishing boat in the Cordova harbor, "was named after the Austrian archduke, whose assassination by a Serbian separatist sparked World War I." I'd never heard of Franz Ferdinand before, but over the course of six days, I discovered this was but one tiny detail of Peter's rich and well-integrated knowledge of history, art, science, and engineering. I inquired more about his commitment to farming. He liked owning his own business and his responsibility of stewarding the land. He valued carrying on a tradition of farming that had been in his family since his Norwegian immigrant ancestors had settled the land in 1864. In our final hours together, driving through the Alaska Range, Peter asked if we could enter into a dating relationship. It was the question I'd been hoping for, but I didn't know how his commitment to a farm would fit into my life as an Arctic researcher. I said yes. I regarded Peter too much to hastily close any doors.

For a year and a half, we corresponded by old-fashioned letters, email, and phone calls. Peter made frequent trips from Minnesota to Fairbanks to help me remodel my new house. He was not the least bit intimidated by new projects. In fact, he thrived on them. He rebuilt the bathrooms, ripped out old flooring, and created tiled surfaces so beautiful I'd have thought this was what he did for a living. Peter's analytical mind and thoughtful approach to interpersonal relationships was also very useful when it came to solving problems that arose in my newborn research group at the university. Peter frequently helped me and my students with field and lab work. When we knew each other well enough to be honest, I told him about my dad and Sandy being initially displeased at his joining us in Cordova. "Did I do the wrong thing?" I asked. "How else could we have gotten to know each other?"

"We would have found a way," Peter answered, revealing his love for treating people right above all things.

My trips to Minnesota were more seldom, but they provided insights into the work ethic and kindness of his parents and into the huge gamut of responsibilities and talents Peter had at the farmstead. On the twelfth night of Christmas in 2008, seated around the living room sofas with his mother and father, Peter handed me a box containing his great-grandfather's late 1800s wedding ring, the ring of the first farmer to take a wife on their farmland, as a proposal to me of marriage.

I knew in my heart that Peter was the man I was destined to marry, and yet I could not for the life of me reconcile what that would mean for the practical reality of his Minnesota farm and my Arctic science. I accepted the proposal but spent the next six months worrying about the repercussions. Did he and his parents really think I was ever going to come live on that farm in Minnesota? It wasn't that the farm was not beautiful. The grounds were full of flower and vegetable gardens, fruit orchards, Black Angus cattle, lovely architecture, and well-maintained historical barns. I'd never had more peaceful nights of sleep than in the second-story guest bedroom of his mother's attractive home. In summer, I awoke to a symphony of birds, their songs carried in with the sweet breeze past the white lace curtains to my pillow. In winter, I watched through the same curtains as soft snow fell on the barns and Norway spruce trees. I savored the serenity of the farm on my visits. But visiting was one thing, and the notion of making any address on Earth a permanent home was another. The latter was like a ball and chain. Besides, Minnesota was flat. Beyond the picturesque farmyard, corn and soybean fields stretched to every horizon. There was almost no wilderness to speak of in southern Minnesota, and it was a long way away from permafrost, glaciers, and Arctic lakes.

Four months later I was leading a five-week lake-coring expedition on the remote snow- and ice-covered northern Seward Peninsula. While winter camping with a small group of students and colleagues, I had plenty of time by myself to wonder what getting married would mean for my freedom (see center insert photo 17). All my life I'd been on my own, whether I had liked it or not. I had made all of my own decisions to travel, study, and research anywhere in the world I'd been inclined to go. Would getting married and having a family put a stop to that? I couldn't bear the thought of it. Peter said I would have to trust him. I knew he was right. And while my human brain could not reconcile the disparities of Midwest farming and Arctic adventure, I knew that this was a journey I was going to have to let God lead me on.

...

MY PLANE LANDED in Minneapolis from Frankfurt, Germany, on June 16, four days before the wedding. For several weeks, I'd been working with one of my students at the Max Planck Institute for Terrestrial Microbiology. We were opening our eight-foot-long cores of Northern Seward Peninsula lake sediment inside an oxygen-free glove box. Inside the anaerobic box we scooped sections of cores into 300-milliliter glass bottles. After capping the bottles, my student would remain longer in Germany to monitor how much methane the sediment layers would produce. Halfway across the world, Peter and his parents were planting crops, painting barns and fences, and preparing the farm for a big wedding.

Rural Minnesotans don't hire other people to do their yard work. They do it themselves. The flower beds were immaculate. The vegetable garden was garnished with daisies and cosmos. The cedar walls and decks of the large, central farmhouse were newly stained. Fresh yellow paint coated support beams on the grain bins, and the trim on all the barns and sheds was a clean, bright white. Two large tents and a stage had been erected on the lawn in front of the house. Tables beneath the tents were set for two hundred. I arrived in time to help cut flower stems in the basement for bouquets of pink and white roses.

On June 20, 2009, Peter and I were married in his family's nineteenth-century Norwegian Lutheran country church two miles across the fields from the farm. A tractor-drawn hay wagon carried the bride and groom, along with a multitude of children, my three sisters, and our closest friends, from the church back to the farm. The wedding and reception that followed were joyful. I knew twenty-two of the two hundred guests. I was thankful to have these friends and family—who had come from as far away as Germany, Mexico, Nevada, and Alaska—see at least for once in their lives the place I would come to call home.

When Peter's best man got on the stage with his guitar and began a song he'd written about the meaning of my name, the words struck a guilty chord in my heart. "Katey, pure as the snow." Most people at the wedding did not know the irony. Tears trickled down my cheeks and fell on the white satin wedding dress I'd borrowed from my best girlfriend in Fairbanks. I was not pure. Almost exactly a year before the wedding, my anxiety about my past had come to a head. My relationship with Peter had become more serious by that point, and the friend, whose wedding dress I would eventually wear at my own wedding, implored me to tell him about my past before he proposed marriage.

That moment, ten months before my wedding day, had been a painful crash between my old life and my hopeful new one. For months I dreaded confessing to Peter unvirtuous acts I'd committed as a young woman, during the years when I'd turned my back on God and during the years I had climbed the slippery slope back to my faith. I chose to tell him in early September 2008, on the last day of a four-week expedition we'd taken with students and colleagues to study thermokarst lakes from a base camp of tents erected on the Northern Seward Peninsula tundra wilderness. Peter and I had spent weeks digging soil pits together around the edges of lakes, sketching and sampling the layers of soil. Peter worked diligently with the shovel, asking all the while the value of data we were collecting. It wasn't so much that he doubted we had good reasons to do the work but rather that he wanted to learn what the effort would yield since this hard work with soils obviously had nothing to do with bushels of grain. Making confessions about my personal life didn't feel right at the soil pits, so I waited.

On the lakes, it too seemed easier to keep talking about science. Peter filtered water samples while I sampled bubble traps. I explained to him that we needed to find out if the ages of the methane in the bubbles were the same as the ages of the permafrost soils surrounding the lakes. The water samples he was collecting would help determine if the thawing permafrost soils were releasing nutrients, such as nitrogen and phosphorus, that can act as fertilizers of plants and algae that grow in the lakes. Nutrient runoff was something Peter cared a lot about. The nutrient fertilizers he applied to his crops were expensive, but the plants needed them to grow, and he did not want any nutrients to escape his fields and run off into the ditches, creeks, and rivers, where water quality was a major issue in the lower forty-eight states.

We hiked across miles of tundra back to camp each day. Peter hauled a heavy frame pack loaded with science gear and samples (see center insert photo 12). He would look mighty good, I thought, with little babies topping a pack like that, if one day we should get married and have a family. But then I felt the burden of the confession I still needed to make to Peter before any hope of a future together.

In camp we ate our supper with the others. I put samples away in temperature-controlled storage, a foam-covered pit we'd dug into the frozen ground. I organized field gear for the next day and then crawled into my tent and sleeping bag. There I listened as Peter bathed in the frigid water of the lake outside our tents (a farmer always takes a bath before bed) and to the loons calling to one another across the smooth water's surface at dusk. Peter dried himself off and then crawled into his tent, next to mine. Since we were not married, he did not want to give others the false impression that we were sleeping together. I had never encountered such prudence in field camp before, and it was a far cry from the giggles we could hear coming from a tent on the other side of camp, where one researcher's tent companion was a student, not his wife. But who was I to judge? I knew that my own past was branded in black, and somehow I needed to tell Peter I was not the innocent girl he thought I was.

When the final day of field camp came, our team of nine people bundled personal gear, the cooking tent, the science and electronics tent, and all of our field gear and samples into a large pile in the marsh grasses next to the lakeshore. We watched as our bush pilot circled the huge lake in his floatplane and then came streaming across the water's surface, sending tall wings of spray behind the plane like a giant swan. It took a full day to ferry our team and gear back and forth from Cape Espenberg across Kotzebue Sound to the town of Kotzebue, which would be the first step in Peter's journey back to Minnesota. If my confession was going to lead to a breakup, I thought it would be better in Kotzebue than amidst the field campaign on the tundra.

I told Peter what I needed to over bowls of noodles at the Bayside Chinese Restaurant overlooking the breakwater on Kotzebue's front street. He listened with large, intense eyes but didn't say anything in response. After paying for supper, I returned alone to the Bureau of Land Management's bunkhouse, where our field team was staying while waiting for our commercial flight to Fairbanks. For hours I paced a sterile room while Peter, somewhere outside, roamed the dusty dirt roads and neighborhood alleyways of Kotzebue. Not able to stand it any longer, I went outside to look for him, but to no avail. Hours later, long after dark, Peter returned to the bunkhouse, still silent. Without speaking, he went to bed. Without speaking, we boarded the morning flight to Fairbanks. Without speaking, he took a taxi with me to my house on Chena Ridge. He washed his clothes and repacked his bags for the red-eye flight that night to Minnesota. When the preparations were done and he sat down, I broke the silence.

"I knew I needed to tell you the truth, but I feared your knowing would mean you are no longer interested in a relationship with me. Is this the case?"

"Not necessarily," Peter said. "God is a God of forgiveness. I need more time to think, and there is a friend I'd like to talk with back home."

Peter never told me which friend he talked with and I never asked. To this day I do not know. Whether or not it was Eric, singing his guitar song on the stage at our wedding, I knew that I had been forgiven. Marriage to Peter was my second chance in life. It was a chance for a stable, successful family life that spoke to a deep inner longing rooted in the failure of my own parents' marriage and my unstable childhood. I was blessed to marry Peter, a man who shared my faith in God as well as my love for nature, my pursuit of knowledge, and my commitment to scientific research. On the other hand, the marriage would require some degree of settling down on his Minnesota farm. This variable still seemed to pose a threat to my adventurous scientific work. Could this too be reconciled? I believed with the help of God it could. We would find out all too soon.

...

SIX WEEKS LATER, Peter and I were bound for Siberia. I might be married now, but so far my life in the field had not been interrupted by this new domestic sphere. With a newly funded NSF grant, I was eager to pursue a science question that had burned in my mind since being a graduate student. I also wanted Peter to meet Sergey and come to love this corner of the world as I did.

Our biplane to Cherskii flew much lower to the ground than previous jets I'd taken. Clouds overhead filtered the sunlight, giving us clear views of the extensive larch forest dotted with thermokarst lakes. Steep, muddy banks showed signs of active permafrost thaw and collapsing ground around the lakes. My heart beat as hard as it had on my first trip to Siberia, the land of a million thermokarst lakes, the land I cherished.

I put my hand on Peter's knee. Behind us, six weeks and halfway around the world, was the long Scandinavian table in his parents'

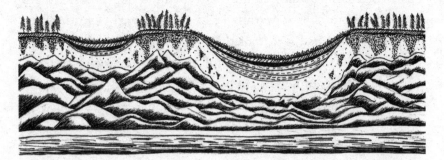

Peat-rich lake sediments in bowl-shaped depressions, the remains of ancient thermokarst lakes as seen in cross section from the river

farmhouse, spread with twenty different kinds of delicious home-made wedding cakes and exquisite flower bouquets. Ahead of us were sixty-foot cliffs of permafrost soils, layers of which we would sample with shovels, spades, and drills; layers of which looked like giant chocolaty ice cream cakes that had been cut in two. Only in Siberia the knife was the river.

As a graduate student six years before, I had motored up the Kolyma River with Sergey. We looked at vertical permafrost cliffs where the meandering river had cut through the landscape, exposing cross sections of ground. At that time, I was sampling crepe-thin layers of ancient lake sediments beneath bowl-shaped depressions that once, thousands of years ago, had been the bottoms of water-filled thermokarst lakes. The lakes had long since drained. Their sediments had refrozen. Then the river had migrated and cut into their basins, exposing the sediment layers in a vertical cliff that looked like the exposed half of a giant layered cake. My goal as a student in 2003 had been to learn the birth dates of the lakes in order to reconstruct what their methane emissions had been long ago when the lakes still held water and their sediments were not frozen. What I had discovered then, and subsequently published in the journal *Science*, was that when these lakes first formed, they emitted vast quantities of methane that contributed to climate warming as Earth emerged from the last Ice Age.[1] But I had seen something else at the time that I never wrote about. In my mind was still imprinted the image of ten-foot-thick layers of mossy, red peat draping down over the crepe-thin lake sediments I originally went to sample (see center insert photo 11). This tantalizing image had stuck in my mind, luring my thoughts for six years to raise money and return to Siberia.

Peat meant carbon. It meant carbon that the mosses and other plants had at one time soaked up out of the atmosphere when they were alive and photosynthesizing. Until now I had focused all of my research efforts on greenhouse gas emissions, but the presence of so much peat implied only one thing: carbon uptake and storage. Was it possible that the plants growing in the lakes were a counteractive force on climate warming? Were they causing climate cooling by removing carbon from the atmosphere and sequestering it as peat in the lake beds?

When our biplane touched down on the runway, the first thing I wanted to do was show Peter my beloved, well-studied Shuchi Lake. The Bible says that when people marry, the two shall become one. But it doesn't say how long it takes for the two to become one. I mistakenly assumed it happened magically at the altar when Peter and I said our public vows. Carrying my old metal oars with Peter through the larch forest along the narrow path to Shuchi Lake, I anticipated the feelings, experiences, and sensations I had had for so many years when I was there by myself as a graduate student. But something was different coming here with my new husband. I was excited and he was not.

We reached the lake together and found my old red boat, up-turned among a patch of dwarf birch and willow bushes. The autumn colors of the foliage no longer seemed the unbearably vibrant magenta, red, and yellow I had remembered. What shocked me most, though, was that where the vegetation had once provided a stable launching point to get the boat out onto the water, I found the bushes were now just islands in a fan of mud oozing toward the lake. At the top of the fan, where the lake margin was steepest, the earth was cracked. Chunks of soil were breaking off, falling into the muddy fan and slowly creeping with the oozing mud toward the lake.

"Wow," I exclaimed to Peter. "This used to be a relatively stable place to launch my boat. Sure, there were signs of soil cracks and thermokarst back then, but the cracks were mostly covered with vegetation. None of this oozing mud was here when I was a student. This is a sign that the ice in the ground surrounding the lake is melting faster now."

Peter remained silent. He was listening and watching. He was also wondering what on earth he would be doing of value for the next four weeks, time that he could easily have filled making important, practical improvements to the farm.

I was eager to get out onto the water to look for the methane bubbling seeps I used to monitor with my umbrella-shaped bubble traps. I sat in the stern and rowed us out across the still, black water. My asymmetric oars once again creaked in their locks. Spurts of bubbles rose here and there in the water.

I could tell from the flat expression on Peter's face that I was the only one who was thrilled. I tried not to let his mood dampen my spirit. I was looking for a particular seep, the seep I'd named Hotspot #4. It was a seep I had monitored year-round as a student, measuring its bubbling rate daily and collecting samples of the gas to study in the lab once per month. No one had bothered with the seep since I'd left, but I knew the seep well. It should have been about sixty feet from the eastern thermokarst margin, where the water was around fifteen feet deep. Suddenly I spotted it! A plastic bottle-buoy still stood erect just below the water's surface next to where the bubbles were continually streaming up. Seven years earlier I had tied the empty bottle to a brick, dropped the brick on the bottom of the lake, and allowed the bottle to bob up toward the surface, marking the location of Hotspot #4.

Dime-size silvery bubbles wobbled up before us from the deep, dark lake bottom. "What I learned from this particular seep, Peter, is that despite the year-round bubbling activity at this location, the methane in the bubbles has a different radiocarbon age at different times of the year. In summer the methane is five thousand years younger than in winter. I think the same ancient permafrost carbon is fueling methane production year-round, deep down at the bottom of the thaw bulb sediments under the lake. But in summer, the surface sediments heat up and add younger carbon to the bubbles of ancient carbon as they pass through the sediments."

Peter was a sharp thinker and was catching on fast, despite his gloomy expression. "But we're here to study that," he said, pointing to a thick bed of mosses and pondweeds growing underwater closer to shore.

"Exactly!" I answered. "Permafrost methane emissions are a positive feedback to climate warming, but plant photosynthesis leading to peat formation and carbon sequestration is a counterweight. It's like the two processes are operating in opposite directions, pulling on either end of the climate change tug-of-war rope."

"Which process is stronger?" Peter asked.

"I don't know. That's what we're here to find out," I answered.

We carried our oars back through the woods to the science station and left them on their shelf in the basement of the old, yellow laboratory building. I took Peter's hand and walked him out to the bluff, to the point where the decorative, wooden church stood overlooking the Panteleha River and the vast Kolyma River floodplain. Fifty feet below us, the Panteleha River's clear water ran swiftly westward. Within sight of where we stood, a branch of the Kolyma River, murky brown from all the sediment it carried, ran into the Panteleha River channel. For miles, these two different waters flowed side by side, unmixed, sharing the same channel. But eventually, as their waters gurgled and flowed over irregularities in the channel's bed and swirled past rough edges of its banks, the two waters merged and became one. By the time they reached the Arctic Ocean seventy-five miles to the north they were completely mixed.

Memories of this landscape poured across my heart and mind. Beside me, Peter stood tall and still. Our hands were clasped, yet I had never felt more disconnected from a person before. Cherskii was my treasure trove. I had desperately wanted Peter to share equally in its riches, but he did not relate. How could he relate? Cherskii meant as much to him as a Minnesota cornfield meant to me. The next day we would be departing on an expedition, traveling along the Kolyma River to the Arctic Ocean to study lake sediment and peat deposits exposed in coastal bluffs. Would Peter and I, like the Panteleha and Kolyma rivers, ever become one?

...

WHEN HE'D MET us at the airport, I'd noticed that Sergey had slightly aged. His hair had grown quite long and he wore it in a ponytail, bound with a pink hair tie borrowed from his granddaughter, Katya. His calculated movements, however, remained those of big-ship leadership. Later, standing outside Sergey's front door, I could hear Sergey and his son, Nikita, shouting loudly inside. I paused to listen before knocking a second time on the doorframe. Somewhere beyond the protective mosquito netting, deep inside Sergey's carpeted living room adorned with trophy skulls of Pleistocene lions and mammoth tusks, the two men were vehemently debating the amount of nitrogen needed to sustain the grassland ecosystem that once had fed the large megafauna during the last Ice Age. I realized that what had changed most strikingly was which little ship had become the center of Sergey's attention. My little ship was now on the periphery of his purview. In the center was Sergey's son, his own flesh and blood. Sergey was set on grooming Nikita, a strong and handsome twenty-five-year-old recent mathematics graduate of Novosibirsk State University, to become an established scientist and the future director of the Northeast Science Station. I was glad for both of them that Nikita's interest in science had grown genuine. Yet part of me was also a little jealous. It was the part of me that wished I was Sergey's offspring, that I could spend a lifetime living and working in Siberia alongside Sergey.

Uneasy about eavesdropping, I knocked a second time on the wooden doorframe, but more loudly. Galya appeared and then disappeared to let Sergey and Nikita know that I was there.

"Ah, good morning, Katya," Sergey said to me as he came to the door, donning his rubber hip waders and a black French beret hat. He stepped outside. "Today we will go to the Arctic Ocean. Are you ready?"

"Yes," I said. "All of our bags are packed." I summoned Peter, Guido Grosse, my contemporary and German co-investigator on the project, and two graduate students from the guesthouse. Loaded with backpacks and cases full of sampling supplies, we followed Sergey, who himself was loaded down with boxes of cigarettes and canned meat. He led the entourage down the steep, rocky bluff path to a large barge moored on the river beneath his house. When the barge wasn't being used as a floating laboratory for scientists, it served as Sergey's personal docking system. We stepped carefully across the barge's deck so as not to disturb long lines of carefully laid out bones. The bones were the remnants of woolly mammoth and other Ice Age mammals, part of a pet project of Sergey and Nikita to combat climate change.[2] Sergey directed us to two thirty-foot boats tied to the back of the barge.

"These boats are for expedition," Sergey explained in broken English to our group of foreign scientists. "Blue boat is for captain. It has good, strong engine. It will also carry Argo."

We looked at the Swedish-built, six-wheeled Argo all-terrain vehicle parked behind the wheelhouse on the blue boat. "I bought Argo specially for your expedition. Argo drives on tundra and swims across rivers," Sergey boasted. "The red boat, we will tow." Sergey pointed to the second hull, which did not have an engine. "I built cabin for you."

We looked at the pretty wooden structure sitting atop the engineless red hull. My heart went out to the little cabin at once. Inside, scientists could sit in comfort while looking out large plate-glass windows. The cabin had a woodstove and kitchen alcove, and Sergey had fastened to the wall a map of the Kolyma River Delta and shorelines of the Chukochi and Krestovsky Capes west of the delta, the places we aimed to go.

"Down there," Sergey said, pointing to the hold beneath the forecastle on the blue boat, "is enough dry wood to last a month, in case anything happens and we get trapped on the Arctic Ocean."

Nikita and Andrei, a sturdy twenty-year-old local Cherskii resident and deckhand for the expedition, came past, delivering crates of vodka to the hold beneath the wheelhouse.

"Mostly, it is not for us," Sergey assured. I knew Sergey was sensitive about people's impressions of his relationship to alcohol. He didn't want Peter and Guido to misunderstand him. He'd too often been accused by angry peer reviewers of being a stereotypical drunk Russian when he attempted to publish ideas that challenged the Western scientific paradigm. They simply didn't understand that Sergey, whose last name translates as "winter," was a man who had lived so long alone in Siberia that his thinking was often outside the Western paradigm. Sergey considered his own scientific work too important to risk at the bottle. In all the years I had known him, I'd seen him drunk only twice.

"This vodka," Sergey explained, "is for negotiations with reindeer herdsmen. They don't have use for money in the tundra."

"Peter, Katya, here is your honeymoon suite," Sergey said as he opened the thick steel hatch leading to the forecastle of the red boat. We placed our bags on one of the two sets of bunks and climbed the ladder back up onto the deck.

The two boats were lashed tightly together at the stern and bow with thick ropes. They would travel down the river side by side. Behind each large boat, small outboard motorboats fastened by painters (bow lines), were intended to ferry us to shore for daily fieldwork.

With everybody on board, I glanced around. Our boat had no guardrails. A slippery deck on rough water could be dangerous, I thought. I saw a handful of life vests on the deck. "Are there enough life vests for everyone?" I asked.

"Foreigners will wear life vests," Sergey answered. "Russians don't need them."

The engine started and we raised our hands to wave goodbye.

Nikita's pretty wife stood close to Galya on the shore. Galya held Nikita's firstborn daughter, Katya, in her arms. Galya's blue sundress flapped in the breeze as the women waved their arms to us, but the faces of both mothers bore a foreboding look of concern.

...

NORTH OF CHERSKII, the main channel of the Kolyma River widens to nearly three miles. From the middle of the channel, I

Thermokarst lake near Cherskii prior to fall freeze-up, September 2003.

Methane bubbles trapped in early winter lake ice, October 2003.

LEFT: Similar to my early investigations with Sergey in Cherskii, here I ignite a pocket of methane trapped in early November lake ice, Fairbanks, Alaska, 2009. This flame is from three weeks' worth of bubbling at one point. Each lake has thousands of points of bubbling and there are millions of lakes in the Arctic that bubble all year round. Most of these methane bubbles escape naturally to the atmosphere, contributing to the greenhouse effect.
Credit: Todd Paris, UAF

RIGHT: Terry Chapin in Sergey Zimov's kitchen, April 2001.
Credit: Mimi Chapin

Northeast Science Station as I first saw it in April 2001. The Zimovs' house *(white, left)*; guest house *(middle, green roof)*, old yellow laboratory *(right)*. In 2002 a new, small white laboratory was added to the center of the station.

Homemade bubble trap in Shuchi Lake, July 2002. The skirt funnels bubbles rising from the lake bottom into the clear plastic bottle with the blue sampling port. Brown, recycled beer bottles are floats.

Sergey considers an actively thawing embankment of yedoma permafrost, 2002. The exposure was initially cut by the Kolyma River. Shiny surfaces in the exposure are massive ice wedges; dull surfaces are soils containing carbon-rich remains of the Ice Age ecosystem.

Sergey Zimov overlooking the Kolyma River floodplain, July 4, 2003.

Our floating base camp on the August 2009 Arctic Ocean expedition to sample frozen lake peat along coastal exposures. Deckhand Andrei *(left)* and Sergey *(right)* await our crew's return from the day's fieldwork on shore.

Simeon *(left)*, me *(middle)*, and Vasya *(right)* at their home in the village of Zhigansk after our ice-bubble fishing field trip to their izba (homestead) in North Siberia, October 2005.

Peter smiles at me on our way to dig a permafrost soil pit on the Northern Seward Peninsula, August 2008.

I am standing under a blanket of thick peat, which was once the bottom of a thermokarst lake in Siberia. This discovery in August 2003 inspired the thermokarst peat expedition to the Arctic Ocean in 2009 that ended in a shipwreck.

After several years of research in Siberia, I turned my attention to methane bubbles in Alaskan lakes too. This launched a career-long effort to measure bubbles in over 300 lakes around the pan-Arctic. Shown here is a methane megaseep in Qalluuraq Lake, Alaska, October 2008. Vigorous bubbling keeps the hole mostly ice-free in winter.

Me (*far left*), Annie, Mom, Christa, and Hanna in Eugene, Oregon, 1983.

My dad and Grandpa Sid, Reno, Nevada, in the early 1990s.

Valera and me at the Kuban River in Krasnodar, 1993.

I contemplate my acceptance of Peter's marriage proposal while conducting fieldwork on the Northern Seward Peninsula, April 2008. Tent camping in bear country with –30°F temperatures required a shotgun and keeping well-dressed at all times of day and night.

Our family enjoys a hike along the Chena River in interior Alaska, September 2017.

Strapped into a stroller, Jorgen could not toddle into dangerous ice holes while accompanying me for fall fieldwork in Alaska, October 2013. The white patches under the dusting of snow are methane bubbles.

Kolyma River Delta

could barely see the western shore. Miniature silhouettes of larch trees dotted the horizon. Moments later the trees were out of view and only a pencil-thin black line marked the edge of land sandwiched between river and sky.

"That must be the transition from forest to tundra," I said to Peter. "The tree line boundary is pretty sharp on the west side of the Kolyma River. We can't expect to see another tree for three weeks now."

The sun had set, painting the sky a brilliant gold along the horizon. I rolled my head back to see where the sky's fiery color faded into deepening shades of nighttime blue. My gaze landed on a bright white, quarter-size moon hanging overhead. Beside us, thirteen moons danced as rippling reflections on the water's surface. My mind was on science. Peter gave me a squeeze and said, "Let's go inside for supper."

Inside the cabin, the jovial chatter of expeditioners diminished as Sergey set down his pocketknife and a freshly opened can of peas among several different forms of dried, salted whitefish and bottled beverages. Sergey exulted, "I found our anchor on a wrecked ship. It is the best anchor in the Kolyma." Excitement splashed across the faces of everyone as we toasted glasses of beer and sparkling water and talked about world politics and culture late into the night. Then everybody retired to bed, except Sergey, who exchanged places with Nikita in the wheelhouse. He would navigate down the river all night.

At dawn, where the Kolyma Delta opens to the Arctic Ocean, we turned west, following the coastline. Nearing the broad shoulder-shaped peninsula of land known as Chukochi Cape, a 120-foot-tall permafrost cliff popped into view. Sunlight glinted off shiny, dark brown patches along the cliff face. I recognized instantly that these shiny patches were the massive wedges of ice embedded in frozen yedoma soils that had formed in the ground over thousands of years during the last Ice Age.

Sergey steered toward shore, cautious not to get too close. Dark mud from thawing soils oozed down the cliff. It dripped and slipped off the icy precipice onto the beach below, where a cornucopia of ancient megafauna bones and driftwood lay scattered among house-size blocks of frozen earth that had crashed to the ground from the disintegrating permafrost wall. "That cliff is not safe," Sergey said, turning back out into deeper water.

We motored further west to where the height of the bluff undulated between high and low levels. The tall banks were yedoma itself—the remnants of the Ice Age ecosystem from the late Pleistocene. Where the bank dipped into bowl-shaped depressions, massive ice wedges were missing. This topography was identical to what I'd seen six years before, as a graduate student with Sergey along the Kolyma River south of Cherskii (see center insert photo 7), except here there were no trees.

"That is an alas!" I shouted and lifted my binoculars as we drew closer to the first depression. "Alas" is derived from a Yakutian term for a drained lake basin. "See the bowl shape of the basin? Long ago it held water. And when it did, the permafrost soils with ice wedges thawed beneath the lake, creating the big depression in the ground."

"Why are some drained lake basins higher in elevation than others?" Peter asked as we motored closer to a highly perched basin, the bottom of which was only about eighty feet above the ocean's surface.

I struggled to focus my binoculars on the outcrop, searching to see if I could make out any dark, peaty layers in the sediments of the high lake basin. "It is a matter of how much ground ice melted beneath the lake," I explained with binoculars still stuck to my face. "If the water-filled lake sat around on the landscape for long enough, then the underlying ice would have enough time to melt, causing deeper and deeper subsidence of the basin. But if the lake was short-lived, or if there wasn't much ice in the ground to start with, then the basin would remain highly perched—just like the one we're approaching!"

My voice had intensified as Sergey motored us toward shore. Finally, I had the sediments in focus and could clearly see beautiful layers in the cross-sectional cut of the lake basin. One layer in particular was very dark. It must've been peat. But this dark, peaty layer was relatively thin and it was under the laminated lake sediments, not on top of them.

"That must be the trash layer," I said, disappointed. I'd been hoping for thicker layers of peat on top of the lake sediments. I handed the binoculars to Peter. "The trash layer is the layer of the contemporary tundra ecosystem—or forest ecosystem, further south—that gets rapidly flooded when the ground subsides and the lake first forms. Do you see how the trash layer extends as a continuous dark line of peat across the entire lake basin? In the forested region, huge tree trunks, branches, and roots protrude from the trash layer. Above the trash layer, there are finely layered, crepe-thin lake sediments. They consist of alternating bands of material that washed into the lake from its surroundings and of aquatic peat formed from mosses, other plants, and organisms growing and dying in the lake."

I was silent for a moment, wondering whether or not to voice my concern. I had expected to find thick layers of peat spilling out over the top of the lake sediments, like the peat I'd seen as a student in the exposures south of Cherskii.

"I don't see much peat here," I finally admitted, deflated and confused. "I wonder if we'll find any thick beds of lake peat at all this far north in the tundra."

In my mind, I started to doubt my initial scientific motivation. I hoped this expedition that cost Guido, me, and the US taxpayers close to a hundred thousand dollars, as well as much thinking time and planning on Sergey's and everyone's part, was not a wasted effort. I didn't give any thought to what Peter had at stake, his own career, his farm. Selfishly, I figured he was lucky to have this chance to come to Siberia with me. But thoughts about Peter quickly flitted from my mind altogether. I was anxious about locating lake-bottom peat so I could find out if thermokarst lakes played an important role in mitigating climate warming.

"Be patient, Katya." Sergey was reading my mind. "We are going to see many alases," he shouted over to me from the open window of the wheelhouse.

It felt good to hear Sergey's reassuring words. I knew that he only got up off his couch if he believed the science question was worthy of effort. This was a question Sergey and I had thought up together. He had told me when I was a student that this was a good science question, that I should not delay pursuing it, otherwise someone else would. Intense competition and drive pulsed through me. This expedition was my chance to be the first to collect these data, to prove to the world that lake peat was important to climate regulation, to continue to prove myself as a competent scientist.

But, good science is honest. It avoids bias. Even though I didn't see much peat through my binoculars, I knew that I had to sample every alas we came to. Each drained lake basin would be a valuable statistical data point, contributing to the mean that would tell the story of whether or not my science idea was right. The eighty-foot-tall frozen alas cliff, undercut by the seacoast, was too steep and dangerous to sample, but I knew we had to find a way.

A little farther on, we saw a place where we could approach the shore in the outboard motorboat. Akin to his father's nature, Nikita patiently waited in the boat, conserving his energy for more logistically draining tasks than taking science samples for foreigners. But unlike his father, he did not light up a cigarette to deter the swarms of mosquitos around his head. He sat tall and proud, watching Guido, the students, Peter, and I zip up our mosquito netting and clamor out of the boat.

Immediately one of the students' rubber boots got sucked into the mud, which was more like quicksand than anything. I suddenly recalled my own first experience getting sucked into the liquefied silt of a yedoma beach with Sergey. I had grabbed for a piece of driftwood and had lain my torso on it to pull myself up out of the ooze. My calves had come free with socks still tucked into my Carhartt pants, but one boot had stayed somewhere down inside the dark mud. "Be careful, Katya," Sergey had instructed me. "You have to move quickly. This is like quicksand."

I gave the same advice now, but the student was already faring better than I had. We shuffled our feet quickly. Without stopping, it was easier to stay on the mud's surface and not to get sucked down. Where we could, we leapt onto protruding bones and driftwood until we reached the base of the cliff, where heaps of sediments were a little drier and more stable.

I found a place without undercutting where I thought we could safely climb the cliff. "What do you think, Guido?" I asked. Guido had a decade of experience working on exposures like these in another region of Siberia where the Germans had based their research on yedoma permafrost since the mid-1990s.[3]

"This is good," he agreed, and started up the cliff with a spade in his hand and a pack on his back.

We followed him. Near the top, the cliff grew steeper. I grabbed for shrub roots as my feet slipped on mud slithering down the icy cliff face. Pulling myself up over the lip of the bluff, I crawled on my belly somewhat relieved to find myself facing dry tundra. The thick scent of Labrador tea filled my nostrils, and my mouth watered when I realized I'd scrambled right up into a patch of ripe, lowbush blueberries. The others were already feasting. I picked a handful of juicy, plump berries and quickly opened my face-net zipper to pop them into my mouth before the swarming mosquitos could invade. Then I stood up to get a good look around. I'd floated in boats through lots of thermokarst lakes before, but this was the first time I'd stood in the empty bottom of what had once been a forty-five-foot-deep yedoma lake. The ancient lake margins rose like the walls of a giant cereal bowl around me on three sides. Behind me was the East Siberian Sea, which had eroded the coastline and eaten into this ancient lake basin. Under my feet, tundra grew on top of the cliff-exposed ancient lake sediments.

Our foreign scientist team worked hard all morning excavating the seasonally thawed sediments from the cliff's face first with spades, then with chisels and spatulas, till we had dust in our throats and eyes, splashes of mud all over our clothes, aching backs, and weary arms. Guido and I hovered close to the exposure, examining and classifying the layers of frozen, carbon-rich mud. There were some paper-thin layers of peat in the sediments, but not as much as I'd expected. Guido hammered chips of frozen sediments out of the permafrost, and Peter placed them in carefully labeled sample bags. I recorded the field notes. By the end of the afternoon, we carried heavy rucksacks back to Nikita's waiting motorboat and then placed them in an electric freezer Sergey had installed in the central hull of the large red passenger boat. Only then did Peter ask a question that made my heart sink.

"If our goal is to quantify carbon, don't we need to collect a specific, known volume of material? Our bags are full of amorphous chunks of frozen muck."

He was right! With our samples we could measure the concentration of carbon but not its bulk density. But I didn't see how we could possibly hope to get uniform shaped samples from a frozen wall of mud. We all saw the chips of icy sediment flying from Guido's chisel and hammer.

"Peter's right," I said, turning to Guido. "You've sampled lots of permafrost carbon before. How do you and your colleagues convert carbon concentration to carbon bulk density?"

"It is hard," he said. "We have to assume a standard bulk density conversion factor. Hardly anybody has ever actually *measured* bulk density."

Peter's brow furrowed and his lips narrowed. He was deep in thought. "Too bad I didn't know more about the sampling process beforehand," he finally said. "I bet a hole saw and drill would have done the trick."

"You mean the little metal cup-shaped attachment with sharp teeth that you use to cut holes for door knobs?" I asked.

"That's it," Peter said. "Instead of chips we would have small, uniform cylindrical pucks of frozen soil."

Thousands of permafrost-cliff samples had been collected by scientists before, in other parts of the Arctic—practically all with chisel and hammer. For the most part, everybody relied on a standard bulk density conversion factor to calculate how much carbon was in the ground. That assumption introduced a lot of error and uncertainty in results. Here a practical farmer was offering a resourceful solution that could revolutionize the power of our data.

My intuition was proving right already. Peter was a smart, problem-solving sort of man who found satisfaction in being useful. I was fortunate to have him for my husband. I saw already that he would be an asset to the science. "If only your toolbox wasn't sitting on the other side of this ocean," I murmured, looking out across the flat blue water that stretched to the horizon.

During the next seven days we sampled twenty-five alases along Chukochi and Krestovsky Capes and a handful of intact yedoma cliffs interspersed between alases. One day we'd split up and Sergey had taken Peter and Guido to a distant set of alases inland. This was an opportunity to demonstrate his latest pride and joy, his swimming ATV, the Argo, which he boasted could cross both tundra and water. The journey required several river crossings. As the Argo plunged into the first river, the current started to spin the vehicle and carry the crew downstream with seemingly no power for successfully crossing. Peter and Guido immediately grabbed some oar ends and began to paddle with all their might from the rear of the Argo. Oblivious to the paddlers behind him, Sergey observed that the Argo was back on course and making steady progress across the river. Pleased, he sat up taller than before and steered as a proud commander, keeping his gaze fixed on the river in front of him. At the next river, Sergey turned around and saw the young men paddling. He shook his finger in disproval. "No, no, no," he said, "Argo swims."

Sergey had been right about being patient: we did find thick peat sequences and were able to gather a good quantity of samples and a fully intact mammoth tusk. It took four people to lift the tusk from the mud and haul it into Nikita's outboard motorboat at the end of the thirteenth day. By the time we drew near the large boats that were our base camp, the sun had set. The seawater was as smooth as glass (see center insert photo 9). From afar we could see Sergey standing calmly with his back against the wooden cabin on the stern of the passenger boat, watching us. The curve of his large, round belly was silhouetted against the white painted walls of the cabin, and on his head was his beret. When we arrived, Sergey helped haul the tusk on board.

Only after the tusk was secured in the central hold of the passengers' hull, alongside our freezer of samples and the spare propane tank, did Sergey tell us the bad news. "Our engine is broken," he reported. "I am waiting starter parts. Come, please, Peter," Sergey said, beckoning Peter toward the engine. I was proud that Sergey thought enough of Peter to confide in him about technical matters. He'd never done that with me. I wanted to be close to Sergey; he was like a father to me in science. Peter was my other half, so Sergey's invitation to him felt like he was drawing us both in closer to himself.

When they returned, Peter explained that in an arrangement called a pony motor, lots of older large engines required a small gasoline engine to get started, instead of using an electric starter motor. The pony motor on our ship had a cracked block. Galya finally had located a pony motor from an old bulldozer near Cherskii, and some Evenki people were on their way to deliver it.

Sergey continued, "As soon as the engine is fixed, we will drive home tonight. A storm is coming."

"Tonight?" I shrieked. I was crushed to hear we'd be leaving so soon. There were still more alases on Krestovsky Cape that we needed to sample in order to have enough statistical power in our data set.

"I afraid," Sergey continued. "If we don't get back into the mouth of the Kolyma River before storm, it can carry our boats far into ocean. We can get trapped in the sea-ice pack and spend winter floating with ice toward Greenland."

During supper the pony motor arrived. A young Evenki boy

Storm over the sea

looked up at me skeptically from a small motorboat as his father and grandfather passed the motor up to Sergey. Nikita strung an electric light bulb to the top of the mast on the captain's boat and turned it on to light the engine house. In the glow of this light, a gray ringed seal with long white whiskers emerged from the water and circled our boat. Its tiny wake formed a V across the water's surface. Then the Evenki family pushed off and motored away into the darkness.

...

SOMETIME IN THE night I awoke to a nightmare. The clanging of pots and pans in the kitchen was not a dream. Real metal clashed with real metal as the hulls of our two boats slammed together, scraped sideways, pulled apart, and crashed together again. Water beat against the portholes. I reached across and grabbed for Peter's arm on the adjacent bunk. He awoke with a start and together we listened as the hulls bashed to the rhythm of a rough sea. It was morning, but the sky was still dark. Through the portholes we saw Sergey, Nikita, and Andrei struggling to loosen the thick ropes that bound the boats together. Their cotton sweat suits and jeans were soaking wet.

"If the hulls continue to bash like this, they'll break each other apart," Peter stated in alarm.

Sergey rapped on the porthole glass, beckoning Peter to come outside. In Sergey's hand was a long sharp knife. He was going to cut the ropes, and he needed Peter to help man the lines. Peter dressed instantly and leapt outside. "Be careful, Peter!" I called to him as I scrambled into my life vest to follow.

Peter was out of sight by the time I climbed the stairs and opened the hatch. I crouched in the doorway and looked around. The slippery wet mid-deck was no place to linger. Water poured off the deck as the boat tipped sideways on a rising wave. As the wave fell, our boat came crashing down on top of the blue captain's boat. Water shot fifteen feet into the air between the boats and rained down on me. I squatted low to keep my balance and then rushed across the open deck to grab the door handle of the cabin.

Safely inside, I braced myself on the table and boxes. I teetered across the cabin floor strewn with spilled dishes and tools to reach the window that looked toward the bow. I needed to find Peter, to make sure he was all right. "Thank God," I exclaimed as I spotted his fluorescent-orange raincoat. He was stationed behind the railings, fastening one of the thick ropes to a cleat.

Inside the cabin, Guido and the students were out of bed and had already donned their rubber boots, rain gear, and life preservers. To everyone's relief, the banging and screaming of metal ceased as the large boats finally separated. But alarm crept into my mind as I watched the captain's boat pull out in front with Sergey, Nikita, and Andrei on board, leaving our boat of inexperienced seafarers to drift behind.

The captain's boat disappeared behind the crest of dark blue waves, but soon a gentle jerk on our boat meant that the seventy feet of line connecting the hulls had suddenly pulled taut. The towrope was our umbilical cord connecting us to the only people who knew how to navigate the wild sea in which we were thrashing. This cord was our lifeline to the only motor strong enough for such rough water. There wasn't a sign of land in any direction. Only the captain's towboat came in and out of sight with the crest and fall of waves.

For hours Peter worked at the bow to keep the ropes taught and out of the propeller beneath the blue captain's boat. I admired my husband for his courage, strength, and competency, which caused Sergey to trust him enough to appoint him a seaman's job.

Inside the cabin, the rest of us looked at one another in fear and disbelief. "If we have to, we could drive the little motorboats to land," I suggested to Guido. We looked out the window but quickly lost heart as the two boats were beating each other up in the storm, wearing their side-seam aluminum thin and shattering the plexiglass windshields.

Hours passed as we traveled in this configuration. When thin rays of afternoon sunshine broke through the clouds, turning the sea from dark blue to green, the waves subsided. Peter came into the cabin soaking wet and exhausted. I handed him a piece of bread spread with Galya's wild blueberry jam.

"I wonder where we are," I said to Peter. "We haven't seen land since yesterday."

Peter didn't answer. He took a bite of blueberry bread and stared wide-eyed at the floor.

As the sea calmed down, Sergey's crew refastened the two hulls side by side to make faster headway. Rumor had it that Nikita had mistakenly driven 90 degrees off course all through the previous night, in the direction of Greenland, not the Kolyma. It was not entirely his fault since Sergey had decided to leave their compass at home on the fireplace mantel. But now Sergey had redirected the boats and was trying to regain position toward the Kolyma River Delta. Only now it was too late to avoid the storm.

As darkness fell, the winds picked up. The waves grew stronger than the night before and sheets of icy rain poured down. The hulls began crashing together again. Sergey and Nikita set the best anchor in the Kolyma. It held but a few minutes and then the cable spool snapped off the boat. The Russian crew scurried about the two hulls turning, shouting, and swearing. Bashing of hulls amplified. The loud Russian argument drifted in and out of earshot between the bangs and screams of scraping metal.

Peter and I sought shelter in the forecastle with the students, but Guido was still in the above-deck cabin. Just as I opened the forecastle hatch to check on him, he stepped out of the cabin onto the slanted, slippery deck separating us. Suddenly, behind him, the cabin door and entire wall to which it was attached tore off the boat and whirled high up into the dark, wet sky. As Guido moved forward toward me, a wave washed over the surface of the deck, sweeping more of the cabin into the sea. I reached out my hand toward Guido and helped him through the hatch.

Guido stood dripping wet in the center of the narrow forecastle floor. "Get into your sleeping bag," I shrieked, horrified that we'd almost lost him, the person who remained behind to make sure everyone else had made it out safely first. I took Guido's sleeping bag from his hands and laid it out on a top bunk. Without speaking or bothering to take off his wet clothes, Guido climbed up onto the bunk and crawled into his bag. He didn't speak again for two days. He just laid there, staring at the ceiling of the tossing hold. The students, Peter, and I sat upright and wide-eyed, listening to the clashing of voices, water, and metal outside.

"How long will the boats hold up to this bashing?" I asked Peter in fear.

"I don't know," he replied.

For hours we lay on our bunks, frightfully listening. Then sometime in the middle of the night, the bashing stopped. The hatch swung open, and a sopping wet Sergey, Nikita, and Andrei descended into the forecastle. "We cut the ropes. The other boat is altogether gone," Nikita said. "We are drifting now."

...

EIGHT PEOPLE CRAMMED into the four-person quarters. Everyone lay down where they could. Nobody uttered a word, but I was sure the same question blared in everyone's mind: What would become of us? We were pitching to and fro on the open ocean in a boat with no engine. I lay for hours on my side with my back against the inner wall of the hull and my chest against Peter's strong, damp body. My heart ached for Peter's parents. They'd tried for more than ten years to have children, and had even given up career opportunities to work for the World Bank in Nepal so that they could remain in the United States to get fertility advice. Finally, at the age of forty, Rachel had become pregnant with Peter. He was their miracle child. He was the apple of their eye. They had never spoiled him but raised him up as a thoughtful, hardworking boy who grew into a man of great integrity. Their prayer had been that Peter would find the right wife and enjoy the happiness of marriage that they had known. How joyful they both had been, sitting on the couch, heads covered in silver hair, the Christmas day when Peter proposed. How happy they had been at our wedding. When I married Peter, I had known that in a way I was taking him away from his parents. But until this moment, I did not think that I would be taking him away forever, to a watery grave.

Our boat was so small, and the sea so vast. All through the night water beat against the porthole glass and washed over the deck. I held Peter's hand, drifting between sleep and prayer. "*Save us, God.*" By morning the waves were smaller. Sergey, Nikita, and Andrei got up and left the forecastle, instructing us to remain inside.

"You can use this for a toilet," Nikita said, pointing to the bucket on the floor between the bunks.

The third night the sea roughened again. We slept, accustomed now to the boat tossing about on large waves, until sometime in the middle of the night we awoke to a bump. Our hull lifted, tilted steeply sideways, and came crashing down loudly on something very hard. Oblivious to the cause of the crash, we held our breath. Again, our boat lifted, tilted, and crashed down. A new pattern had been established. The mysterious crashing was much louder than anything we'd heard since the storm had set in. Each time the boat tilted, the smaller of the two students flew down out of the bunk above us. Each time she flew, Peter extended his arm upward, caught her in midair, and tossed her back into her bed. This exercise went on for hours. The loud, angry crashing sound terrified me. Surely our hull would not stand up to this beating. It had been nearly twenty-four hours since we'd seen the Russians. Why had Sergey not come to us with an explanation or any reassuring words? What had become of them? Were they still alive? Were they even on board?

Peter had the mammoth tusk on his mind. We could hear it beating against the bulkhead, electric freezer and propane tank in the adjacent chamber. At any moment, the tank valve could break off and the tank would come shooting through the thin metal bulkhead to explode in our hold. With dawn's morning light, the tilting and crashing became gentler. We peered through the portholes to see what we could make of the situation. To our great joy, in the dark shadows that linger above Earth before sunrise, we saw land!

...

STANDING ON SOLID ground never felt so good. Who cared if the ground was soggy with floodwaters and driftwood that were rising by the minute? I leapt off the boat without asking then turned around to watch the others follow. To my relief I saw that the Russians were alive and on board. However, it seemed as if Sergey had no desire to speak to us. He was shouting commands to Andrei and Nikita to investigate the damage of the wreck.

With long blond curls now knotted by weeks of wind and dust and twisted into a tangled wreath, I rested my head on Peter's chest. I wondered what he thought of this honeymoon expedition. Raindrops fell, turning cakes of dust on my scalp into muddy streaks that ran down my cheeks. I pulled the hood of my parka up over my head. Wasn't it just six weeks ago that we had been standing before the altar at the little country church in Norseland, Minnesota, committing before hundreds of people to serve each other until death do us part? So far, I hadn't served him very well. Together we stared at the ship's wreckage strewn along the soggy shore in front of us.

Guido's GPS unit showed that our drifting boat had had the good fortune of catching on the northwestern tip of the last island in the Kolyma River Delta. On the other side of the island, which was less than half a mile away, there was nothing but ocean between us and the North Pole. Calling this spit of land an island was generous. The land rose altogether no more than three feet above sea level, and I could easily see water at what seemed eye level on the other side of the shrubs. Driftwood, scattered among willows and soggy tussocks, told us that it was not unusual for this island to be completely flooded. We watched as the storm lifted and carried our wrecked boat higher with each crashing wave up onto the flooding island. I was worried.

"It's only a matter of time before our boat bounces on the rising water across the entire island and is back on its way to Greenland," I told Peter.

"I found the satellite phone," Guido said triumphantly, approaching us with long, intent strides. "We could call the NSF's logistical support company and ask them to send helicopters."

Sergey was furious when I suggested placing the call. "That is stupid, Katya. You will not call. I am the captain," he sputtered.

But I am the leader of the foreign team, I said to myself. *It is my job to make sure the students and all of us safely survive this expedition.* I knew I could not go against Sergey. I felt like the lowly graduate student hugging her knees on the step of his science station as he lectured me so many years ago. Except at that time, his fatherly discipline was a sign that he cared enough to invest his time and emotions in me. But now it seemed as if he didn't care at all. Hadn't three days gone by with not a word to me from Sergey, since the night he'd said we would head for home because a storm was coming? Standing on this soggy shore with the phone dangling at my side, I despised Sergey's pride. He didn't want to admit that we needed help, especially not from American helicopters. What I was not rationally appreciating was Sergey's foresight that involving an American rescue mission would mean the end of opportunities for US-funded research expeditions at his station. I was so ready for this adventure to be over that I could not fathom desiring ever to come back again. But with the passage of time, that would change.

"I'm going to go help Guido work on the cabin," Peter announced. We made our way back to the boat, and he set in with

Guido to find spare plywood to rebuild the one and a half walls that had blown off the cabin in the storm. Nikita climbed down into the hold with the mammoth tusk. Standing in a soupy mess, he passed up my science backpack with a GPS and soggy field notebooks. He passed parts of the freezer that had been completely demolished by the mammoth tusk: first the lid, then the motor, and finally the bashed-up remnants of the freezer box. Fortunately, our samples, still double bagged in Ziplocs, were dry. I gathered them together in my arms and carried them gently to the forecastle bunk.

When I returned, I found Sergey and Andrei in the repaired cabin tearing pages of a book and rolling them with tea leaves for smokes. Sergey took a drag on his homemade cigarette and closed his eyes. Long bags of red skin sagged beneath them on the top of his cheeks. Tendrils of smoke curled around his head. He ignored me. He had said all he was going to say. I had never felt so estranged from this man in my life.

Outside I found Nikita and asked him what the plan was to get us out of here. He said that his father had just used the last juice on their satellite phone to call Galya. The Russian Ministry of Emergency Situations should be here tomorrow.

TWELVE

Gassy Lassie

I, a tattered soul, returned to Fairbanks with Peter. The expedition seemed like a complete failure. In science, convincing arguments require repeatable evidence. Large, statistically powerful data sets are one means to that end, but my rescued peat samples were just a dismal, partial data set. The only way to prove my idea about lakes sequestering atmospheric carbon was to return to Siberia and get more peat samples, but that was about the last thing I could ask Peter to do. Even I dreaded the thought of a return.

What should have now felt safe, inside the confines of my university office perched on a hill and overlooking the boreal spruce forest of Fairbanks, did not. Without a manuscript in the works, I was not living up to my vocation of research professor. I was like a baker with nothing in the oven, a fisher without a net in the sea, or a farmer without seeds in the field. My research goal of the past six years had ended in a wreck. Sergey, my closest mentor, had become estranged. How he could have the audacity to leave his compass at home and allow us to get into the shipwreck with no words of communication, consultation, or comfort along the way was all beyond me. How could I ever trust him with my life, and the lives of my husband and coworkers, again?

I sat at my desk wondering how to pick up the pieces and move on. My eyes wandered to a white sheet of paper tacked to the wall. On it was a simple pencil sketch my grandpa Sid had drawn and mailed to me from Nevada. A stick-figure girl kneeling inside a rubber raft, paddling. Printed below were the words "Methane Queen of the Klondike, the Gassy Lassie."

I was now in the business of Arctic methane, and with this reputation it was not uncommon to receive letters, emails, and phone calls from people all around the North American Arctic bringing to my attention various gassy phenomena that struck them as weird or from media groups wanting to cover the topic. I found myself getting nostalgic for safer expeditions from years past that had focused on methane instead of peat. I recalled one day in July 2007, nearly two years earlier, when the phone rang and opportunities aligned. Melissa Block from National Public Radio was interested in doing a story on Arctic methane and wondered if she could join me on a field trip to the northern coastal plain of Alaska. She had funds for a helicopter, which for me meant an opportunity to investigate an unusually large gas seep located in a remote tundra lake outside the village of Atqasuk. The lake was called Lake Qalluuraq—Lake Q—by the Iñupiat people who lived in the region. I had never been to this lake, but a colleague had tipped me off to it with a blurred photo of frothing bubbles and a report that a seventy-year-old Iñupiat elder said the gas seep had been there ever since she was a child. The photo had poor resolution, but it was enough to convince me that a large quantity of gas was very likely belching out of the lake, and I wanted badly to see it for myself.

A week later I found myself jumping out the open door of a Bell 206L-4 Long Ranger, my Xtratuf boots landing in a thick bed of soggy, wet sphagnum moss. Melissa had been first out of the bird and already had her large fuzzy microphone turned on, holding it in front of me when I jumped to capture every sound detail on our adventure. Flying over the lake before landing, I had seen a disturbance in the lake's surface, caused by what looked like strong gas bubbling. Now my job was to find that spot on the ground. I didn't even have any GPS coordinates to guide me. As I pumped up my rubber boat, I forced my mind to dismiss thoughts of doubt and a fear of failure. *You'll get this, Katey*, I said to myself. I'd learned while playing college volleyball that my chances of scoring were much higher when I visualized a successful attack on the set ball before striking. Doubt only led to failure. Optimism and fortitude had always served me well in science too, but never before had I had someone else's expensive helicopter bill and the precious time of a national newsagent on the line. What if I failed? I couldn't think about that.

I dragged the eight-foot-long inflated white Zodiac over to the lake's edge and set the small outboard motor in place. Melissa climbed into the boat amidst my bubble trap and gas sampling kit. I pushed an oar into the soft, peaty lake edge, nudging us out into deeper water. I hadn't started a motor on my own in years, not since my days of sampling the rivers and lakes near Cherskii. The sound of the propeller churning the water after a few pulls on the cord was a relief. I turned my eyes to the vast silvery lake and wondered which way to steer. I had a general sense of where to go, having sited the location of the seep approximately 150 yards northeast of the helicopter's landing spot. But now, at lake-level elevation, finding a single point of gas bubbling in a lake the size of 250 football fields felt like looking for a needle in a haystack. I set out slowly toward the northeast, searching for any sign of anomaly in the lake's surface caused by bubbles. Waves, the spatter of rain, or even ripples from wind would wash out the texture of the bubbles, making them impossible to spot. Fortunately, this day the lake's surface was flat, like a silver mirror. I motored forward, keeping my eyes aimed in the direction where I thought the bubbles should be. Suddenly I saw them! I let out a victorious whoop and stared, gaping, excited, scared. Of the thousands of methane seeps I had studied, I had never seen anything like this one before. Instead of small percolating bubble events reoccurring every few seconds, in front of me was a three-foot-wide plume of violent gas, a dome of bubbles erupting incessantly at the lake's surface. Hundreds of smaller bubble streams rose around this focal point, contributing to the churning of the water column above a thirty-foot-wide dark hole in the lake bed.

Having located the seep, I suddenly found myself afraid to approach it. Would we get asphyxiated? Would the bubbling cause our boat to sink, and would we vanish like boats in stories of the Bermuda Triangle? There was only one way to find out. Pushing fear aside, I resigned not to leave without getting a sample.

I motored straight toward the bubbles. A slight breeze now spread over the lake, causing frothy bubbles to stream eastward. I motored into the seep from the west, breathing fresh air until I was suddenly positioned right on top of the seep with bubbles erupting beneath the thin skin of my rubber boat floor. Bubbles streamed up around us on all sides. To my great relief, we were still floating! A smell akin to a rotten egg filled my nostrils, but after multiple shallow breaths, I assured myself that I still felt all right.

I quickly cut the motor and dropped an anchor. Grabbing my bubble trap, I flopped over onto my stomach, pulled up my sleeves, and tried to submerge the trap. The bubbling was too strong. Large, forceful bubbles pushed up from beneath, pinning my trap to the lake's surface. There was no way I could get my trap down. Without that I couldn't get a clean gas sample. I grabbed the oar and paddled several feet away from the bubble plume. In quiet water I was able to submerge the trap. Then, still hanging over the side of the boat, holding the submerged trap with one hand and an oar with the other, I made small strokes to maneuver back over the violent seep, but this was to no avail. The forceful gas plume just pushed the boat away. After multiple attempts, my small one-handed paddle strokes finally penetrated the plume. The trap instantly filled with gas, causing the folds of the skirt to bulge up out of the lake. With gas now inside my trap, I could let the boat drift away from the bubbling plume to calm water, where I could more easily transfer the gas from my trap into small glass bottles. I held up the first sealed bottle full of transparent gas to examine it. Sunlight reflected off small drops of water clinging to the outside of the bottle, but inside was pure gas. An excellent sample!

"You seem really excited," Melissa said, extending her microphone boom toward me. "Where do you think this gas is coming from?"

In the wonder of this frothing foam of bubbly seep, I had forgotten all about my company. My mind raced to answer her question. This definitely was not the same type of methane I was accustomed to sampling in thermokarst lakes. True, the lake itself was a thermokarst lake, formed by thawing permafrost, but this gas plume was just too big and too strong to be coming from decomposing detritus. Maybe there was a pile of mammoth carcasses decaying in the sediments, but that seemed unlikely.

"I can say what I think it is *not*," I answered, almost too excited to utter my thoughts out loud about what the gas might actually be. "It's not the normal source of methane that I see in most Arctic lakes. This gas smells like rotten eggs, but methane is odorless, so there must be sulfur in it too." Then I let it all out: "I'm betting this is geologic methane—ancient methane from coal or other hydrocarbon deposits that has been trapped for a long time and is now escaping through the permafrost. If I'm right, this could be a really big deal. It means that permafrost thaw not only feeds organic matter to microbes that make methane but also opens up conduits for deeper pools of trapped fossil hydrocarbons to leak into the atmosphere."

During the flight home I twisted my body to look through the chopper window, down into the green and blue waters of thousands of lakes below. A breeze sifted the water's surface of most lakes, but the lakes were shallow, and in many I could see through the water to the lake beds. My eyes scoured the sandy and muddy beds of these lakes, searching for any signs of more gas seeps. I saw nothing like the gas plume at Lake Q, but I did see dark, round cavities in the bottoms of other lakes. Perhaps those dark cavities were pockmark signs of other large, bygone gas seeps. A myriad of questions started to race through my mind. How many methane megaseeps were out here in the Arctic? Once formed, how long did one of these megaseeps last? How was it that this gas was streaming up through nine hundred feet of permafrost? Wasn't permafrost an icy cap that impeded the flow of natural gas? How much methane was actually coming out of Lake Q? Could huge plumes of gas like this be useful to people?

Now, in September 2009 sitting at my office desk in Fairbanks, my eyes lingered on my grandfather's stick-figure drawing. So much had changed since that 2007 expedition to Lake Q, notably Peter. I remembered the January 2008 return to that lake with him: he had been visiting me in Fairbanks during our courting days—a year before we got engaged—and he had agreed to help me find out how much gas was escaping Lake Q. Clearly my small, flimsy bubble traps were useless on a methane megaseep this immense, but Peter put his mind to work. At a time of year when the sun doesn't rise above the horizon and every patch of our skin was covered by thick beaver furs and goose down to protect us from a –50°F air temperature, Peter's idea worked. We used the natural ice cover, which was less than two inches thick over the seep (a striking contrast to five feet of ice in the rest of the lake) to trap the gas. We laid a strip of tarp over the ice surface to seal any cracks, poured water on it to freeze it to the rest of the ice sheet and then drilled a hole through the tarp and ice to focus the gas stream above the seep. Finally, we installed a pipe with a flow

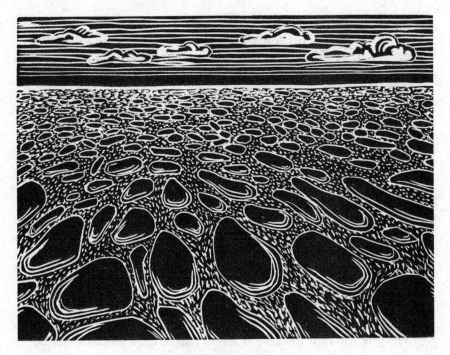

Thousands of lakes

meter to measure the gas escape rate. This approach showed that this single seep emitted enough natural gas to meet the energy needs of six Iñupiat households. The problem was that it was too far away from where people lived to be of much use. We left asking ourselves, how many such megaseeps were out there, relentlessly gushing this potent greenhouse gas into the atmosphere?

Here I was, nearly a year later, in the wake of the most challenging expedition of my career. Grandpa's drawing on the wall reminded me that there were other scientific questions I wanted to pursue. It would do no good to wallow in the failure of our peat carbon expedition to Siberia. It was time to refocus my efforts and pick up the pieces to help solve the Arctic's geologic methane mysteries.

Peter was game to help me.

"I have an idea of how to locate other methane megaseeps!" I exclaimed that evening in our kitchen with newfound purpose. "I am confident that if I could get up in the air right after the lakes freeze and are covered with a dusting of snow, I'll be able to locate more megaseeps."

Peter raised an eyebrow, so I explained: "This approach worked for me in Siberia, only there I was mapping smaller methane seeps. Fast bubbling creates open holes in ice. Have you ever seen how people use air bubbling to keep their boat slips ice-free in winter?" Peter nodded, so I continued. "Methane megaseeps should do the same thing—create great big open holes in lake ice (see center insert photo 13). If I could get up in the air in early winter, right after the lakes first freeze, then I could look down on thousands of lakes across the landscape and locate the black, bubbly holes. They conspicuously contrast against white, snow-covered lake ice." The eagerness in my voice faded as my thought train inched forward. "The problem is, it will take at least a year to get a proposal funded to pay for all those flights and fieldwork."

I shut up. After several months of marriage, I was becoming accustomed to waiting in silence while Peter thought. Growing up an only child on a rural farm, Peter had always had all the time he needed to think before speaking. Now it was just a matter of me being patient long enough to listen to him. Rarely did his responses disappoint.

"If this is exciting and important science, Katey," Peter finally said, "let's not wait for a grant. Let's fund it ourselves."

Fund it ourselves? I had never dreamed of paying for science out of pocket. Peter's entrepreneur mindset twirled with my academic persuasion, forming a single cord. It was true that with money he'd saved from farming, Peter had paid off the debt on my Fairbanks house when we got married. Now with my professor's salary, we could afford a handful of charter flights that would be necessary to fly a north–south transect in Alaska during early winter to map methane megaseeps. Besides, I also had some discretionary funds at the university that could be directed toward charter flights and field trips.

...

IN THE FALL of 2009, two weeks after our honeymoon shipwreck in Siberia, Peter and I each began new projects: I started my self-funded research of studying methane megaseeps and Peter began a master's degree program at UAF to study his farm soils. His parents agreed to manage the farm, largely in our absence for several years, while Peter worked on his degree. Since they weren't getting any younger, Peter reasoned that now would be a better time than later for him to be absent from the farm and to support my work in Alaska. His seventy-two-year-old dad woke cheerfully each morning to drive semitrucks of grain to the Twin Cities, feed cattle, and remove snow from pig barns after blizzards. However, to ease their load, we made extended trips to Minnesota at spring planting and fall harvest. Trapped with my laptop in a dark little house at the back corner of the farm, where Peter's grandmother had lived out her days, I found myself increasingly depressed at the daunting notion that I would someday have to accept Minnesota as home. At the farm, I slouched in the shadow of Peter's mother, who maintained the beautiful large central farmhouse and meticulous gardens. I much preferred Alaska, where I was head of a burgeoning research group, where I met regularly with students and colleagues to fan the synergistic flames of research, and where I could leave campus at the end of the day with Peter to the privacy of our warm and pretty little house on top of Chena Ridge. Peter and I had love, but would it threaten other pieces of my life? For the time being, our lives were in Alaska. I did my best to block out the concerns of a future in Minnesota, but somewhere in a dark, back corner of my mind, they crouched down and waited.

Peter's thesis research goal was to analyze long-term data sets collected on his own prairie farm soils by a University of Minnesota professor. After a cancer diagnosis, this professor had asked Peter if he'd like to take over the analysis. This was a fusion of opportunities to keep the professor involved on Peter's committee and let Peter develop analytical skills in line with his genuine interest in the data sets pertaining to his own farm management decisions. One thing about farmers is that they have trained themselves to sit for sixteen hours a day in their tractors, covering thousands of acres of fields multiple times over. Peter unconsciously applied his tractor discipline to graduate work. After completing his coursework and in between helping me with fieldwork and statistics, he spent the summers of 2010 and 2011 sitting in a tiny loft office in our Fairbanks house analyzing data and writing papers. He averaged a manuscript a month for two consecutive summers, and with six papers submitted to journals by the end of his second summer and a seventh paper in the works, I suggested he talk with his committee about changing the program to a PhD degree. A master's degree requires one paper, a PhD three papers, but in less than two years Peter had already satisfied double the PhD requirement and was writing papers faster than any academic professor I knew. "You just have to keep at it," Peter said.

Peter had never had the personal aspiration to get a graduate degree. He had always been content to teach himself things for the enjoyment of learning and the usefulness of knowledge. Now, acquiring skills in statistical and geospatial analyses under the umbrella of a graduate program seemed like a useful endeavor to him for the future of his farming. I soon found his new analytical expertise a tremendous boon to my own research too. No longer would I have to comb the hallways at UAF to find collaborators willing to assist with statistical programing and geographic information system analysis. Now these skills were located in my own husband, under my very own roof. "A day without data analysis is like a day without sunshine," Peter said enthusiastically each time I brought a new data set to his desk.

My idea for mapping methane megaseeps worked. In the beginning it seemed incredibly freeing to go up in a small airplane and fly anywhere I wanted across the state of Alaska to look for open holes in early winter lake ice. Peter and I flew the northern Alaska coastal plain, the state's major mountain ranges and large river valleys that separated them, and the southern rainforest region where glaciers spilled into the sea. I sat in the front of the plane, searching for holes and documenting them with my GPS unit, while Peter sat behind me faithfully photographing the lakes, even after sharp turns caused him to lose his breakfast on more than one flight. Back

Methane megaseep

in our home office, I pored over the photographs and GPS coordinates, quickly realizing the challenge of ground truthing seeps I had seen from the plane in inaccessible locations. In order to confirm that the open holes were actually caused by methane bubbling, I had to somehow get to these holes with my feet on the ground, peer into them, and affirm bubbling. I also had to measure the gas flow rates and sample the gas for geochemical analysis.

"Be careful," Peter cautioned. We both knew the consequences of falling through the ice in the remote Arctic. I skittered vigilantly across the ice of unfamiliar lakes toward the open holes, one foot dragging an inflated raft and one hand holding an ice spear. When I got close enough to the holes, I chipped away the dangerously thin shelf ice around the seeps, lay down in my raft, and peered into the watery blackness. Concerns about my life and the world faded from my mind. Beneath my nose, streams of silver bubbles percolated up from the darkness, broke at the water's surface, and cast hemispheres of bubbles out like a bouquet in all directions to the edges of the ice hole. I was lost in wonder at what phenomenon lay deep beneath the sediments, sending me these bouquets. Once again I was intimately connected with the mystery of nature. In those moments, little else seemed to matter.

THIRTEEN

Brooks Range

Sitting in the dentist's chair one day, Peter described to the hygienist a recent aerial survey we'd done over a particularly large lake nestled in a valley that was long ago carved out by glaciers in the Brooks Range. The hygienist put down her sonic brush and looked Peter square in the face. "We've seen those lake-ice holes before!" she exclaimed. "My husband and I have a floatplane and a cabin at that lake. We always thought it was swarming blackfish that kept the ice open. You think it's gas? Why don't you and Katey come out this summer to stay with us and do your fieldwork?"

This was an invitation we could not turn down, especially as we mixed self-funded and grant-funded research. We'd seen holes not only in this large glacial lake but also in several of the smaller lakes upstream and downstream from it. A dry shelter and sturdy canoe would go a long way toward facilitating fieldwork in this mountain range above the Arctic Circle.

On an early September day in 2010 Peter and I took to the air in a floatplane heading north out of Bettles. Below us the tundra was a patchwork of autumn colors, interrupted only by islands of sparse green spruce trees and silver lakes. The southern foothills of the Brooks Range jutted up in front us. We flew low beneath the clouds up a gentle U-shaped valley. Our pilot landed on the largest lake in the valley, close to where two small cabins were nestled on the western shore.

Snuggling down in our warm sleeping bags that night, Peter said prayers for our fieldwork. I was thankful to share faith with my husband, a faith that it was God's hand that opened to meet our needs. Peter asked for calm weather conditions so that we could discern methane bubbles at the lake's surface. We didn't expect God to give us everything we asked for, but we trusted in his love and omniscience of what was best for us. Peter and I woke in the morning to rain falling on the lake. I went down to the lake's edge and watched as raindrops made intersecting circles disturbing the top of the water.

Concerned, I turned back to Peter and exclaimed, "How will we ever find methane seeps in this rain? Raindrops obscure our ability to see bubbles!"

"Let's take the canoe and scout out the small lake south of here," Peter suggested, with faith stronger than mine. We loaded our canoe with science gear, donned our Helly Hansen rainsuits, and set off to the outlet river at the southern end of the immense lake, around which only the bases of fog-enshrouded mountain flanks were now visible. We paddled all morning down a gently flowing river of black water, tanned by tundra soils steeped in precipitation. Despite the raindrops, the river's surface was calm, reflecting the colorful shrubs that lined the riverbank among sparse spruce-tree roots, which were losing their footing to thawing permafrost along the meandering river channel. Paddles dipping and the electric pitch of raindrops on water were all we heard. Otherwise, the silence was so pervasive, I was sure that if a bear popped out of the willows next to us, I'd hear it before I saw it. But mostly I was anxious about coming this far to have our fieldwork spoiled by rain.

As the gentle river opened into a small lake, I realized suddenly that a GPS unit was not necessary to locate the open holes I'd seen in our aerial survey. My eyes immediately homed in on a two-foot-wide plume of bubbling. Rain had not obscured the bubbles after all. Although the gas seepage was not as strong as what we'd seen at Lake Q, large pockmarks gouged the sediments beneath these bubbling plumes too. Before nightfall, we mapped hundreds of gas streams in the little lake, paddled back up the river, and found additional huge seeps in the southern bay of the large lake where we'd started. Arriving at the cabin long after dark, tired, wet, and famished, I was ecstatic at our success, but it was Peter who remembered to give thanks to God. I went to bed that night exhilarated by our discoveries. Our fieldwork was proving that the open holes we'd seen in winter lake ice from an airplane marked immense, mysterious gas seeps, the kind no one had ever documented before in the Arctic. I couldn't wait for morning to come, to behold a new scientific adventure at the far northern end of this immense lake. What I didn't know was that we wouldn't be out there alone.

...

TO GET TO the northern lobe of the lake, we had to cross several miles of open water hundreds of feet deep. I was grateful when we reached our destination, where, despite more rain, the water was shallower and large bubbles were visibly rising through the underwater vegetation in the bay. Peter did his best to hold the boat steady and pass me supplies while I worked leaning over the side of the boat with my head under my raincoat hood, intensely focused on measuring bubbling rates and water depths, taking field notes, and collecting samples. Suddenly, we heard a voice call out.

"Watchya doing over there?" It was the voice of a female. We spun around and saw a thin woman with long stringy dark hair in a canoe about a hundred feet away.

I was completely caught off guard. Other than the hygienist's cabin, now four miles southwest of us, we hadn't seen a single sign of settlement. Where had this woman come from?

"Uh, we're sampling these gas seeps," I said, a little shakily. The woman didn't seem very happy to see us there.

"Gas seeps?" she asked. "Oh, we had a big explosion out here once."

"Yes, these bubbles are probably methane gas," I answered, regaining my confidence and composure. "We haven't gotten around to flame testing them, but we're just about to do that. Would you like to come help? I'll need an extra set of hands anyway."

The woman paddled over to where I had one hand gripping the bubble trap's gas-filled plastic bottle that was now bobbing out of the lake. The clear plastic skirt attached to the bottle was partially underwater and full of gas too. The skirt held so much gas that I needed a paddle to push its bottom metal rim down beneath the water's surface to prevent gas from spilling out.

"If I open the valve on this bottle, would you be willing to light a match to it?" I asked the woman. "If we get fire, we'll know it's methane."

The woman agreed and reached out to take the matches from Peter.

An invisible stream of gas whooshed out through the stopcock as I cracked opened the valve. The woman lit a match. "Put it into the hissing gas stream," I instructed. She did so, and immediately a four-inch yellow-and-orange flame sprung up.

"Ha, haaa!" I exclaimed, with a big smile on my face. "We have methane!"

The woman's mouth parted into an awe-filled smile too.

"You can feel the flame if you'd like to," I told her, "but you'll have to be quick. The fire is starting to melt the stopcock."

The woman reached out and swept her fingers through the flame. Then I dunked the bottle underwater to extinguish the fire.

The woman turned her dark round eyes first on me and then on Peter. Except for rosy cheekbones, her skin was pale. She was a Caucasian woman, dressed in rugged mountain clothes.

"I'm Katey," I said, "and this is my husband, Peter. We are researchers from Fairbanks. We flew over this lake last winter and saw holes in the ice where these gas plumes are. So we wanted to come and sample them now to find out what the gas is. We've been doing this same kind of work in lakes all over Alaska."

"I'm Heather," the woman said. "We always thought those were just air pockets."

"Do you live near here?" I asked Heather, looking around and still not seeing any sign of where she could have come from.

"Sort of," she replied. "Come on with me, out of the rain. I'll give you some hot tea."

We followed the woman to a rocky shoal where a creek emptied into the large glacial lake. We pulled our boat up into the bushes next to her canoe and pushed through the yellow willow shrubs behind her as she made her way through a thin stand of spruce trees into a mossy clearing. In front of us stood a one-story log house with an earthen roof and several rough-cut outbuildings, all connected by well-worn footpaths. Everything looked as if it had been made a century before out of nothing but hand tools.

The inside of the house had the same appearance. Little light streamed into the wooden interior. On the walls hung hand-carved wooden bowls and coats made out of animal skins.

Heather poured water from a kettle on the cast-iron woodstove into several metal mugs and set them down on the thick wood table next to a box of tea and some oatmeal cookies.

"I'm the only one who comes here now," Heather said. "I'm trying to keep the place up. I'm working on replacing those logs beneath the roof just now."

"You're doing all of this by yourself?" I asked in disbelief.

"My whole family used to live here," Heather explained. "My mom and dad came out to live here in the 1960s. They wanted to get away from civilization. They raised my brother here."

I couldn't believe it. This woman was voicing the realization of my lifelong, childhood dream to live with a lake in the wilderness.

"They made almost everything by hand," she continued, "and for the most part they lived off the land, hunting and fishing and gathering berries. They went out onto the lake ice to get water from those air holes."

"If only they would have tried lighting the gas on fire," I interrupted. "There is a ton of methane seeping up out there. They could have found a way to collect it and they'd have had their own unending supply of natural gas."

Heather seemed interested in this but then looked down at her cup of tea before continuing her story. "My brother drowned when I was six months old," she said, looking back up at us. "It was September. He stood up in the canoe out there in the middle of the lake. The canoe flipped. No one could get to him in time. My parents took it pretty hard, especially my dad. Three years later, same time of year, my dad walked into a floatplane propeller and was killed."

Heather had spoken matter-of-factly, but I knew that only deep longing could stir her to be out here alone.

"My baby sister was five months old when Dad died. Mom took my sister and me down to California to live. We made trips here with her as often as we could, but now I'm the only one who comes."

A fierce, earnest, and nearly fearful expression came across Heather's face as she looked at us. "We don't really want anyone to know where our place is. Dad and Mom made a film about their lives out here, and ever since that time, people have been trying to find us. We want to keep this place to ourselves."

I understood that. In front of us was a competent and skilled young woman, trapped between two worlds, doing what she could with what seemed like superhuman strength to preserve the cornerstone of her family's former way of life. I had always longed for the wilderness life she described. I too had felt trapped between worlds, living part-time in Siberia, part-time in Alaska, and now part-time in Minnesota, but never really calling any one place home. This woman needed love and friendship, not gawking bystanders. I gave her my email address and agreed to help her with more information about the methane once we had worked up the data.

Several months later, Heather wrote to me. She was interested in the methane. My calculations suggested that, altogether, the seeps in the bay near her cabin were releasing around nine thousand dollars a year of diesel-equivalent gas. This wasn't the first time I was connecting my research on large-scale methane dynamics to help people at the individual and community scales. Peter and I had been working for several years with high schoolers in Cordova to convert cafeteria food scraps into biogas using cold-adapted microbes from lake sediments in stand-alone digesters. We'd also traveled to South Africa and Botswana on National Geographic Society–funded research to help initiate camp-scale biogas digesters there. I wished there was a straightforward way for Heather—like the Iñupiat people of the Arctic Coastal Plain and many other families and communities who live scattered around the vast Arctic landscape—to capture and use this methane. I admired Heather's commitment to her family's wilderness way of life, and I believed that with some ingenuity, of the sort people who live off the land tend to have, there was a solution.

FOURTEEN

Dangerous Ice

People die every year after falling through thin lake ice. For the first decade of my methane career, surviving my own surprise submersions in ice-covered water gave me a false sense of invincibility. Working near hidden, snow-covered holes in ice was a risk I had to take to get my job done. Usually, after unexpectedly plunging into water, it was only a matter of seconds before I was able to find ice strong enough to pull myself back out. But one February day in 2009, while crossing a glacial lake in Southcentral Alaska ahead of two students, I unknowingly stepped into a snow-covered methane pocket and plunged straight down into the icy water. The first student intuitively stepped forward to pull me out, but the thin ice and thick layer of snow gave way beneath her too. We were two girls thrashing about in ice water. Cold water seeped through our winter gear, soaking us to the bone. The weight of my soggy snow boots pulled me down, and I wished I had bothered to wear a life vest. I reached for the ice shelf to try to pull myself up onto solid ice, but that ice only gave way too, crumbling together with thick chunks of snow into the lake, creating islands of slush around us. Fortunately, the second student had decided to remain back. He put down his pack, took out a rope, and tossed it to us, pulling us out onto thicker ice one at a time.

Wisdom comes from experience and listening to others. My students and I were not the only victims of dangerous methane holes. One man, the father of my Iñupiat polar bear guard on one expedition to the Arctic Coastal Plain, unexpectedly plunged through lake ice on his snow machine. Climbing out wet, he froze to death alone on the barren tundra. At one of my study lakes near Fairbanks, a resident dog musher living in a cabin without electricity used a methane hole to get water all winter. One cold April day, when the snow was exceptionally thick and deceiving, this man fell through the ice at his own hole. "I was surprised at how warm the water felt at first," he told me, "but within minutes I felt the heat draining from my body. The ice was so wet and slippery that I almost didn't make it back out." As he told me the story, he pulled a screwdriver out of his pocket. "Now I never go to get water without this," he said, shaking the tool next to his scraggly beard.

In the first few years of our marriage, Peter was nervous about my work on lake ice. I figured it was because he had grown up on a farm and had little firsthand experience around water himself. He only read about the bad incidents in the newspaper. It didn't take long, however, for Peter to lead me out into lake ice situations I myself dreaded.

...

IN 2010, PETER accompanied me to Greenland after the fall crop harvest. It was late November, and it seemed as if we were the only field scientists in Kangerlussuaq traversing the grassy tundra at that time of year. Autumn sunshine bathed the tips of tundra grasses in a golden light as they protruded up through winter's first snow. Wind had blown snow off most of the lake ice, so we did not need to shovel to look for ice-trapped bubbles and dangerous open holes. The opportunity for data collection was at our fingertips. For the first few days our only concern was staying out of the way of defensive herds of musk ox as we hiked many miles between lakes through tundra covered in feces. But as we approached the Greenland ice sheet, we became increasingly tempted to descend from the tops of steep rocky moraines down into the basins of proglacial lakes abutting the ice sheet. The moraines consisted of bare rock with no or little vegetation. Relict icebergs stranded high up on the moraines next to where we stood indicated that earlier in the summer the water level had been more than a hundred feet higher in the lake basins than it was now. I looked at Peter, wondering if we should descend into the proglacial lakes to investigate bubbles.

"What if part of the ice sheet were to collapse on us?" I asked. "After all, the base of the ice sheet is not frozen. It is lubricated with liquid water, causing the glaciers to ebb and flow as a moving river of ice." Neither of us had one iota of experience with this kind of environment.

"Nothing ventured, nothing gained," Peter said, repeating the phrase my eighty-year-old grandpa Sid always used when he accelerated out into an intersection, unsure of the traffic signal indication. "Besides, as the Danish Maritime Authority officer said at the airport, 'Most people die in their beds.'"

In that moment, I loved my husband's sense of adventure. I was glad I was not in this alone.

The first proglacial lake we explored was a long skinny one that snaked around the finger of a glacier. It was surrounded on three sides by twenty-foot walls of blue ice, from which skull-size rocks were cascading down and spilling out across the lake's frozen surface. We shuffled our feet across the slippery lake ice as tendrils of blowing snow swirled around our ankles and made us dizzy. The wind blew at our backs, propelling us fast across the ice. As I glided and slid, I could see that the ice was full of methane pockets, and yet there wasn't a sign of soil or vegetation around!

"Where could all this gas be coming from?" I shouted through the wind to Peter as we came to a stop around an ice wall that offered some protection from the strong gusts. "I don't see any sign of organic matter that would fuel methane production in this rocky glacier."

"Maybe it isn't methane," Peter said. "Let's check."

Peter knelt down and pulled out a knife and some matches. Using his body as a wind block, he jabbed the blade into one of the methane pockets and waited while I lit the match. Whoosh! A brilliant yellow torch of fire leapt two feet into the air, burning so strongly that it started to melt the ice around it. We watched the flame's reflection in the smooth ice surface as it burned down and diminished a few seconds later.

"Well, that's one way to warm up," I said, putting my chapped hands back inside my gloves to escape the biting wind.

Days were short and each hour was precious. We surveyed bubbles on the lake until the afternoon sun set. A brilliant full moon rose over the Greenland ice sheet, giving us light to find the flags we had used to mark ice bubbles in the daylight. To know the chemical composition of the gas, we needed to set traps to collect bubble gas overnight. Peter speared through the thick ice, making two-foot-wide semicircular holes over points where streams of bubbles were rising from the lake bed. I submerged my traps in these holes and dangled them beneath the ice from metal rods set across the holes. It was good we'd brought the rods along, because nothing akin to the larch branches I'd used in Cheskii was anywhere to be seen. These trusty old traps, designed nearly a decade earlier from Siberian refuse, would work for us, collecting bubbles while we slept. Before leaving, Peter opened one more hole. I squatted down next to it and plunged my Hydrolab sensor into the water. It would take some time for the sensor to stabilize so we could record the concentrations of salts and ions, pH and turbidity levels,

Aurora borealis

in three-foot intervals through the deep water column. I switched off my headlamp to wait. To my surprise, the sky was ablaze with streaks of light.

"The aurora!" I exclaimed as wisps of bright green light whipped across the sky overhead, first this way and then that. It moved so fast, with so much energy, I thought I could almost hear the sound of its wind. Peter shuffled his feet, cold after so many hours of standing on ice. "Lake bottom at seventy-five feet," I said when I felt my sensor finally hit something hard.

"Great! Shut off your headlamp," Peter directed. "It's a beautiful night to walk home by moonlight."

...

HAVING SURVIVED THE first lake abutting the ice sheet, we decided to ratchet up our adventure and explore a larger proglacial lake the next day. The risk of an unstable ice sheet wasn't as daunting since most of this lake was surrounded by grounded rocks instead of tall walls of ice. As I scooted out across the ice, I realized that my shuffling boots were sweeping clean streaks on the ice. I was surprised to see that the ice on the lake's surface was obscured by a layer of fine dust and sand; these were the windblown sediments carried from the glacial outwash plains and deposited wherever the wind was blowing. It took a few minutes to adjust my eyes to search for bubbles in these conditions, but I found them soon enough.

"Peter, come look at this!" I called out. Five feet in front of me a tremendous, circular, white pocket of gas extended more than twenty feet in diameter. The bubble was larger than three Peters laying end to end. "I've never seen such a large gas pocket!" I exclaimed.

Peter came over. "Yeah, I found one just like it over there," he said, pointing behind him.

"I wonder what kind of a fire this will make?" I asked, my eyes bulging with anticipation.

Peter jammed the ice spear into the pocket while I readied the match. No flame! We tried another pocket but found the same result.

"Wow, okay, so this is one of those spots with a leaky ice crust," I said, getting over my initial disappointment. "The gas must pressurize inside that white ice bubble and then fizz out through microcracks on its own. I've seen smaller versions of that in thermokarst lakes near Fairbanks. The strange thing is, this lake doesn't seem to have any small pockets," I said, glancing around. "Only great big ones."

Minutes later, I found a pile of sand trapped by a crusty white methane ice bubble. Upon closer examination, I realized the middle of the ice bubble had a hole in it and I was peering straight down into the lake's milky green glacial meltwater.

"Peter, can I have the spear?" I called out with excitement. Peter brought over the spear, and I chipped easily away at the thin crust of sandy ice, exposing a six-foot-wide hole in which large bubbles erupted up through the murky green water. I looked around and realized that wherever we saw small sand piles across the lake there were large open holes beneath cardboard-thin ice crusts. In some places the bubbling was so strong that there was no ice at all.

"These holes would be death traps if we were out here at night and fell in one," I remarked, horrified at the thought of our moonlit hikes back to camp.

I was utterly perplexed. Where was this gas coming from? I could see no sign of permafrost soil degradation around these rocky glacial lake margins to fuel methane production. The water was green because of the dust-particle minerals, not plants.

"Is it possible we've uncovered geologic methane megaseeps here in Greenland as well?" I asked Peter.

"This is one of the neatest experiences of my life," Peter answered. "I have a real sense of discovery out here."

His passionate comment caught me off guard. It had seemed before as if he accompanied me on field trips out of a sense of spousal duty, but now, for the first time, I felt as if we were genuinely, full-heartily in this together.

The sun sunk close to the horizon. Peter and I summited another moraine and looked down into the largest, deepest ice-jammed lake basin we had come across yet. In front of it stood a 150-foot-tall vertical wall of ice that extended more than a quarter of a mile along the lake. Behind the wall was 500 miles of the Greenland ice sheet, which for all we knew would groan and shift and push that headwall forward. Ten inches of lake ice was nothing against a 150-foot-tall moving glacier. Icebergs formed frozen islands in the lake, evidence that my fears were valid. I didn't want to go down onto the lake. Especially not with darkness approaching. There was a lot I was willing to put on the line for methane, but this was too much to ask.

"Please, Peter, I don't want to go. And if we go down there at all, please, let's come back tomorrow," I pleaded. "For all we know, there could be more big holes in the ice, and we wouldn't be able to see them in the dark. Besides, I have a gut feeling that the ice wall is not stable. Standing in front of it, if it calves, could be the end of us."

"We are too far out to come back tomorrow," Peter said. "We're here now. Let's go get the job done."

Peter started down the moraine. I stood watching, afraid to follow but terrified to let him go alone.

We reached the lake's surface as the sun went down behind the Greenland ice sheet. Nausea gripped my throat as I walked out across the lake ice. I wasn't sure if my shaking was due to the cold

Greenland ice

wind seeping through my jacket or the fear I felt following Peter farther across the lake toward the deadly ice wall.

There were methane gas holes, sure enough. Big ones. I knew what my job was. If I wanted to get off this lake, then I needed to sample the gas as quickly as I could. I dropped to my knees, unstrapped a bubble trap from my pack, strung it on a pole, removed my gloves, and dunked the trap with bare hands into the ice-cold water. I watched in the twilight as bubbles streamed into the collection bottle, filling it in seconds. I shook the water from my stinging hands and thrust them into my armpits to warm up before labeling the sample bottles. There was no way I was going to stick around to measure a Hydrolab profile on this lake. I watched Peter wander off in the falling darkness. Minutes later he came back and said he had found an even bigger hole. "I've got samples from this hole already," I said. "I've had enough."

"Okay, let's go," Peter agreed. We packed up our samples and supplies, shuffled off the ice, and scrambled up the moraine. Just as we reached the top, we heard a deafening clap. It was the bang of the ice wall crashing down onto the lake, shattering the lake ice where we had stood. The sound echoed from the rocky hills around us. My heart sank in my chest. I had to look down at my feet to be sure we were on solid ground. I didn't say, "I told you so," and Peter didn't say, "See, we got our samples." We were both just thankful we were safe.

Three days later, on a jet somewhere high above the Atlantic Ocean between Greenland and Scandinavia, Peter took my hand. "Katey," he said, "I think it is time that we start thinking about having children."

Part IV

A PIECE OF CLAY

Then I went down to the potter's house, and there he was, making something at the wheel. And the vessel that he made of clay was marred in the hand of the potter; so he made it again into another vessel, as it seemed good to the potter to make.

Jeremiah 18:3–4, NKJV

FIFTEEN

Back to the Kolyma

I couldn't believe my husband was suggesting we add babies to our lives. This was a bomb more unexpected than the collapsing ice shelf. Ever since I was a little girl I wanted to grow up and someday have children. But someday was still a long way off. At thirty-four, it hadn't yet dawned on me that I should be grown up. I was still too busy living for science and, honestly, for myself. If marriage hadn't challenged this enough already, by causing me to take time out of the day to cook sit-down meals and to follow Peter to his Minnesota farm for a significant fraction of each year, what would babies do to my still self-seeking goals?

Prudence whispered to my heart that I should listen to Peter. If I did not, I might end up like my mother, who was in the agonizing throes of another painful divorce, sleeping again at her parents' house. I had failed in childhood to treat my mother with respect. I had been willful and wild, and from the time when I moved out at age twelve, I'd gone through life putting my tremendous and forceful energy into doing what I wanted. As an adult, I had read that God asks us to respect authority, not contingent on the authority acting deservingly, but for our own good (adapted from Rom. 13). For my good. I didn't have to look back across my past, but simply into my character. I was no better off now for the disrespect I'd shown my mother. I had set out on my own at the age of twelve because I had been determined to do better than she had. But in my pride, I had failed God, my mother, and myself. Now I had a chance to change course. Because of his wisdom and self-control, Peter was easier to respect, but honoring this request had huge implications.

Start having babies right now? There was still so much more science and adventure I wanted to pursue. There were papers I needed to write. My conscience and conviction told me that protesting Peter's request would be selfish and wrong. So I bit my tongue and agreed. Nonetheless, I clung to the redeeming thought that babies weren't something you could get immediately. They were something that took time to acquire, at least nine months, and time was what I needed in order to finalize a few more data sets.

My first priority was to fill in the gaps of my incomplete Siberian peat data set. By the time I obtained a new grant in the summer of 2011, I was three months pregnant. Peter and I both knew this was my chance to complete the long-awaited data set because peat expeditions in Siberia were not something you took a baby along for. When we boarded our flight to Cherskii, my belly was swollen enough that I could no longer button my Carhartt field pants. I was not yet tuned into the world of maternity clothes, and I figured they probably were not field-hearty anyway. A rubber band wound through the buttonhole would have to do the job for this expedition.

Sergey met us at the airport with a big, bushy-bearded embrace. Offenses of the shipwreck were forgiven, but not forgotten or mentioned. Instead, Sergey pulled out the red carpet, slowing down in the Land Rover to avoid potholes and bumps on the frost-heaved road to his station. I'd never seen him drive so carefully before. Without a word, I knew he was demonstrating his

Russian reverence for protecting the safety of pregnant women and their babies.

At first Sergey was apprehensive about taking me on another expedition, given my pregnant state, but I did not relent. Perhaps he too recalled the time, ten years earlier, when he sat on a crate outside his white house smoking a cigarette, watching me cut plastic for my bubble traps. "What will become of you, Katya?" he had asked, looking at me out of the corner of his eye as he exhaled a cloud of smoke. "Will you someday get married and have children and abandon science? That is the nature of most women, you know. Hormones are a powerful force." At that time, I didn't imagine that I'd be one of those women. Now it appeared Sergey was willing to stick with me to find out.

This time Sergey took us south up the Kolyma River in the opposite direction of the Arctic Ocean. There were plenty of alases along rivers in the boreal forest zone too, and nobody wanted to risk another storm at sea. We slept and ate on board the boat at night and went ashore to sample each day. Now it was I, not Sergey, who took the risks. Sampling meant crawling across sloppy, muddy faces of ice beneath dripping permafrost overhangs that could at any moment detach from the cliff, dashing down on me and the baby in my womb. But providence was with us. Over the course of two weeks, my scalp filled with the fine silt of dried, thawed yedoma as fast as the freezer on Sergey's repaired boat filled with samples, hundreds of samples from twenty-five more alases. We doubled the number of alases we'd studied two years earlier in 2009. Finally I had a large enough sample size to draw meaningful conclusions about the role of thermokarst lakes in absorbing atmospheric carbon dioxide.

Before returning to Alaska, Peter and I spent long days in the field station laboratory. Trained by my postdoc, I peered into petri dishes through a microscope, discerning and recording whether the peat carbon in the alas sediments formed from aquatic organisms once living in the lake or from terrestrial plants once living on land. Peter labored away on the adjacent lab bench, scooping small volumes of the peat and sediment samples into empty, pre-weighed white ceramic crucibles, which he then combusted and re-weighed to determine precisely how much organic matter was in the material. He was highly satisfied that his idea to use a hole saw had worked. We were generating what would become the largest data set of precisely measured carbon bulk density for thawing permafrost that existed.

We worked around the clock to get quantitative data from our samples. I knew from previous experience that it could take up to a year to get samples out of Russia through customs. I could not afford that much waiting. With the ever-growing baby in my womb, every day counted. It would be far better to leave Russia with actual data in hand. And data, excellent data, is what we got. But putting the pieces together to form the complete picture of lake carbon balance on the scale of climate change would require time. It was a project I would undertake in the confines of a dark Minnesota farmhouse, bouncing a baby on my lap, with myself an ever-evolving lump of clay in the potter's hand.

A Sleeping Giant

At nine months pregnant, I was frantically writing on our findings from Greenland and Alaska, where Peter and I had discovered methane megaseeps. This was a story about an unsuspected giant sleeping in the ground. The giant was stretching and yawning and burping and attracting attention in ways I could not ignore. The giant was methane itself. Not frozen organic matter in permafrost that had the potential to be transformed into methane by microbes upon thaw, but actual physical methane molecules that had accumulated in and beneath permafrost from leaky hydrocarbon reservoirs, such as ancient sedimentary basins, coal beds, and oil fields. No one knows exactly how large the giant reservoir of methane is, lying in the ground beneath Arctic permafrost and glaciers, but scientists estimate it is several orders of magnitude larger than all the methane already present in Earth's atmosphere.[1] Arousal of this giant by ice melt in the Arctic could unleash some of this trapped methane with profound impacts on Earth's climate. The data Peter and I had collected appeared to be the first evidence of widespread geologic methane seepage along boundaries of thawing permafrost and melting glaciers.[2] It was important to get our results out into the peer-reviewed literature.

My fear was that the arrival of a baby would hinder our publication. While most women I knew bemoaned the final weeks of their pregnancies, desperate to relieve the strain on their own bodies, I was content to swell up like a beach ball. I was thankful that the baby was still inside me, taking its time. It was already two weeks past the due date, but this was two more weeks in which I could pour myself into writing.

My water broke while I was working on my paper. And when it did, I stepped into a whole new existence. Jorgen was born February 7, 2012, days after the thermometer at the gas station below Chena Ridge read –50°F and my belly was nearly as large as the orange bouncy ball on which I sat at our home office desk, furiously writing my manuscript.

The manuscript took a back seat in my life. Something far more important, far more alive, and far more vulnerable had entered my arms and heart. Little Jorgen, beautiful little Jorgen, was so perfect and plump and trusting when he first lay across my breast and looked into my face with his small, dark eyes, taking in all the light and warmth and love his mother and father could give. But all too quickly he began to waste away. The nurses and doctors threatened to put our baby on formula if I could not successfully feed him by breast. I was determined to breastfeed because studies have shown that a mother's milk is the greatest nourishment a baby can have. I tried to nurse him around the clock, day and night, and yet each day he lost more weight. My manuscript faded into oblivion. In my arms was a precious life that depended on me. A love so pure and selfless, that I had never known could possibly come from me, now poured into this baby, the one who had become the center of all my attention and energy.

Equally concerned, yet confident in his experience with piglets, Peter stacked pillows four feet high on our bed to position my breast at his eye level. He tried to get Jorgen to latch on, but it was no use.

Jorgen tried to eat, but my breast only swelled with milk that did not go into his mouth. Jorgen was starving. He screamed for food when Peter picked him up and carried his dwindling flesh swaddled in blankets out into the living room to the warmth of the wood-stove. Baby Jorgen thrust his neck forward, trying to latch on to Peter's nose, sucking desperately. Since the beginning of time mothers have suckled their young. We were determined to get to the bottom of the problem. We bundled Jorgen up in the car and drove him down the hill in the ice fog to the public health center.

A talented lactation nurse stuck her fingers into Jorgen's cheeks while he attempted to nurse. In so doing, she determined that Jorgen was unable to create suction on his own because he was tongue-tied. The nurse said the only solution would be to cut his frenulum, a tissue that connected the underside of his tongue to the floor of his mouth. This simple operation would release his tongue so that it could perform normal suction. The only pediatrician in town willing to do this surgery assured us the frenectomy would be a quick procedure and a relatively fast healing process. Immediately Jorgen was able to suckle, and Peter and I watched with glad hearts as our baby regained his strength. Soon he was a perfectly happy baby, content to sleep on our laps as we resumed academic work.

People get an addictive drive to chase after all kinds of things. For me, each discovery of a large, new methane megaseep gave me a tremendous rise. This was a complicated addiction as a new parent since I was no longer able to travel on a whim to remote parts of the world to study those seeps. It was a newfound inner conflict between my old life and my new one, but I was determined to find a way to make it work.

In October 2012, when Jorgen was eight months old, he and I left Peter harvesting crops in Minnesota to conduct early winter fieldwork in Alaska. This would become our norm as a family, bouncing between Minnesota farming and Alaska permafrost research. Martin Stuefer, a colleague at the University of Alaska, had a two-seat, single-engine Piper Super Cub monoplane in which he offered to fly me to survey for methane megaseeps near the Yukon River region north of Fairbanks. This region intersects the continuous-discontinuous permafrost boundary in Alaska and would fill in a geographical gap in my study. Martin said with his tiny plane we could even land on a gravel bar to look at the ice holes up close if we wanted to.

I placed Jorgen into the arms of a woman from church. Jorgen had never had a bottle before, and I wasn't about to let her give him one, at least not for the next six hours. I calculated I had that much time to work and return before his next breastfeeding. I'd done innumerable flights in small aircraft before, but something felt different this time. It wasn't that this plane was much smaller than most I'd been in but rather that for the first time there was much more on the line should something happen and I not return. A lump moved down my throat as I passed Jorgen over to the other woman and then climbed into the seat behind Martin and fastened my belt.

Soon enough we were airborne. The little bundle of baby-soft blond curls shrunk to a tiny dot on the runway as we flew away. The thrill of methane hunting was upon me. Nearly every lake and slough along the Yukon River was dotted with open holes in ice. What a huge spike this would create on my manuscript map of megaseep methane occurrences in Alaska! If only we could find a place to land and sample. We circled numerous times, looking for a safe gravel bar that would give us access to a lake, but none were to be found. Finally, we settled on a gravel bar adjacent to a frozen slough dotted with round holes. Martin brought the plane in for a smooth landing and we climbed out, grateful to stretch our legs. I was shocked at how cold it was on the Yukon River. Martin didn't want to leave the plane for long due to concerns about the weather turning bad, so I grabbed my backpack and we ran together down the gravel bar in the direction of the ice holes. With almost a year of no practice, my field skills were dull. When the snow-covered sand ended and slough ice began, I found myself uncertain about the safety of venturing out. I could see holes in the ice a hundred feet away. I thought I perceived a bubbling disturbance at the dark water's surface in the holes, but to be sure, I'd have to get closer. Yet this was a river system with flowing water where the ice thickness was unpredictable. If something happened to Martin, there would be no one to fly the plane.

We crept cautiously out onto the ice. If only I'd had forethought to bring an ice spear along. Then we could check the thickness as we went. But time was of the essence and there was no spear. The crowbar Martin fetched from the plane was too dull to effectively penetrate the ice. The ice began to crack as we stepped farther out, close enough that I could discern bubbles rising through the open ice holes. But was it methane? I knew that on rare occasions the bubbles were nitrogen, not methane. Since nitrogen is not flammable, a sure and quick way to tell was to do a flame test. Martin agreed to help me. His crowbar would be useful for penetrating the white pockets of ice-trapped gas around us. He raised the bar in the air and let it fall, piercing a gas bubble as I lit a match. When Martin pulled the bar out of the ice, a gigantic fireball rushed out of the hole straight toward him. Instantly his down coat, eyebrows, and hat were burned. Martin staggered backward, away from the flaming hole. We were both flabbergasted. The entire front of his polyester jacket had melted; the down fibers were burnt black. This was the coat his wife had bought him for Christmas. It was also the only thing he had with him now to keep him warm in these extreme conditions. Realizing the need to return to the plane, I failed to collect a gas sample. Without that sample, I would never know for sure what kind of methane was seeping out of the Yukon River lakes and sloughs.

Our field tests had taken longer than Martin liked. We needed to reach Fairbanks before dark, and clouds were already starting to settle into the White Mountains. Martin took off from the gravel bar and flew east, looking for a clear pass through the mountains. We scanned the sky with apprehension. Patches of blue sky shrunk as fog thickened, its tendrils creeping down into the valleys.

"It's okay. I've been through here many times before," Martin tried to assure me. But as the fog closed in around us, I lost sight of the snow-covered mountain flanks that I knew were all around. Martin spiraled the plane upward and then down, trying to avoid icing on the wings. There was nothing but white outside my window. I lost all sense of direction. For a moment the fog thinned and we both saw that we were headed straight for the rocks in a mountain. Martin jerked the control wheel and rudder, and the plane shot instantly upward. We avoided the mountain this time. Martin was going to risk the icing-up on the plane by flying higher. At least there would be a better chance of surviving this than crashing into the side of a mountain.

I was sick with fear in the back seat. *Please help us make it out of these mountains alive*, I prayed. Airplane accidents are twice as common in Alaska as the rest of the country, and we were getting a firsthand look at why. A clear flight can quickly turn into a nightmare of clouds and precipitation as pilots navigate tricky mountain ranges, glaciers, and meandering rivers. Finally, to our great relief, we pulled out of the clouds and into a clear blue early evening sky. The tree-covered hills surrounding Fairbanks were in front of us and the socked-in White Mountains behind.

"Now you have to help me look out for other aircraft," Martin said. A collision in the sky seemed far less risky than what we'd just been through, but I took his request seriously. As our plane set down on the runway, I knew this was the last time I'd chase methane in a small plane, at least for a long time. With a baby, the risk just wasn't worth it. I would have to be content to wrap up the geologic methane megaseep paper with the data I already had. I thanked Martin for piloting. "I owe you a new coat," I said as I climbed into my car and sped off to find and feed my little one. On the ground, now in control of speed and direction, it seemed I couldn't get to Jorgen fast enough.

A chubby little hand reached out toward the sound of my voice as I came through the door. I drew him to myself and felt complete. In this moment, there was nothing else I wanted. If I was going to make the juggle of my scientific career and mothering work, I would need to find a way to pursue science without taking as many physical risks.

SEVENTEEN

Farm Wife

Seven months later, in May 2013, we were in Minnesota. After Jorgen was born, Peter finished his PhD at UAF and became president of Anthony Farms Inc. This was not a decision we'd made lightly. We both knew that for a family to work, one person's job would have to take priority. Peter put the option out on the table that he could let go of farming and take a wage job in Fairbanks so that my academic career could lead. In my heart of hearts, I would have loved for us to be happy in that scenario, so that we could stay in the Arctic and I could keep at my work full-bore. But I knew it was a scenario in which none of us would ultimately flourish. I also knew that I needed to learn to live out the scripture "Not my will, but yours, be done" (Luke 22:42, NKJV). In Alaska, I was so utterly absorbed in long hours of field-work on local lakes and academic meetings, that Peter and Jorgen took a back seat and I seldom saw them. Peter didn't mind taking over most of the domestic responsibilities in Fairbanks, and he loved baking bread, cross-country skiing, and bonding deeply with Jorgen. However, we were aware that such a lifestyle would work only in short intervals. Peter needed the challenge and responsibility of farming. And I knew that staying in Alaska would foster my lifelong pattern of unhealthy relationships, whereby my academic work was in the driver's seat running over the people closest to me.

Peter was not like me. Regarding productivity, he got more done in a day than I did, even before Jorgen had entered our lives, but his commitment to work never hurt people. He put his work aside when people were near him. He was able to be fully present with them and enjoy them. He was stable, calm, optimistic, and cheerful. To remain that way, I knew he needed to follow his calling to productive work, and his calling was to farm.

For Peter, productivity did not imply work outside the home. He had a vision that the farm could be a wonderful way to raise a family, with everyone involved. The question still remained: what would be my role? Certainly, between my academic commitments, cooking meals, and taking care of a baby, I didn't have time to spend on tractors. Besides, it was not obvious to me which aspect of modern farming would actually fit my skills and interests. I had no mechanical experience, no interest in sixteen-hour shifts driving farm vehicles across flat fields, and there was already a different crew hired to take care of the needs of thirty thousand pigs. I could help in the garden and bring meals out to eat with Peter in the field. For the time being, that would be my role.

Our decision to let the farm take the lead limited our time in Alaska. We would still make extended trips to Alaska as a family for my fieldwork, but when at the farm, I would run my lab group remotely out of the Minnesota farmhouse. In this scenario I also would make occasional short trips to Fairbanks with Jorgen to meet with students and conduct fieldwork (see center insert photo 19). Neither Peter nor I liked it when I took Jorgen up to Alaska alone, leaving Peter in Minnesota. It certainly felt more right to be all together. And so I found myself much of the year in Minnesota, trapped in his grandmother's dark little house in the back corner of the farm, where I was expected to be content with writing papers in between raising a toddler and helping Peter's mother process tons of tomatoes, string beans, eggplants, apples, and other farm foods for winter. What I hadn't yet realized was that this was a place where God planned to lead me through a new kind of adventure—not into exotic wilderness lakes but into the dark inner corners of my heart.

...

ONE PARTICULAR EARLY May day my feet pounded on gravel as I ran down our farm's driveway, pushing an empty baby stroller toward County Road 15. I was late to retrieve Jorgen from a stay-at-home South African woman, our neighbor a mile down the road, who took care of him three mornings a week. Out in the open air with my body moving hard, my mind could start to relax from calculations I'd been focused on all morning. Yet a part of me didn't want to relax or leave that mental space in which I could block out real life. It was easier to remain in my calculations, where I didn't have to confront my feelings.

Behind me was Peter's family farm. From his parents' central farmhouse—an attractive wooden structure with a single plane roof that descended toward the north and ceiling-to-floor windows that looked out to the east, west, and south—a person had a panoramic view of the apple orchard, flower gardens, and the large gravel circular driveway that skirted the grain storage bins, the cattle yard with fifty-three Black Angus and Hereford steers, the big old brown barn with a red clay tile roof and silo, the machine sheds, the tractors, and two other old barns.

In a rock garden near the farmyard's entrance, an obsolete one-bottom plow, flagpole, and a tall wooden post stood together. On the post were engraved the names of the original homestead farmers and their descendants, my husband's lineage. My name had been

Our Minnesota farm

engraved next to Peter's nearly four years earlier, when we were married. Jorgen's name now marked the sixth generation.

Tall Norway spruce trees bordered the farm, shielding it from fierce north winds in winter and giving shade and depth of scenery in summer. Beyond the spruce trees, blossoming wild plum, and lilac bushes lay the green pasture. A long row of large, round hay bales stretched off through the pasture toward the southeast. It extended beyond the shade of an old cottonwood tree and the farm's woodpile. The line of bales marked the route the original county road took 150 years ago, when the region was first settled by Norwegian and Swedish immigrants.

Out on the modern road where I was running, dark, flat fields of freshly planted corn and soybeans stretched to the horizon. A few wooded islands still stood erect among the patchwork of fields, marking the locations of slightly higher-elevation ground, places where trees could survive in the once marshy, lake-prairie landscape that had been wilderness before people moved in to tame, drain, and cultivate it.

The no-longer-wild landscape was for me an eyesore. To escape the emotional pain, I squinted and imagined the features to be something they were not. White clouds on the northern horizon became mountains, and for a moment I could daydream that I would go there. In my mind, clear streams would tumble out of the rocky passes, opening into lush, soggy meadows where the cotton grass would dance among the mosses and lichens. If I could get there, I would walk for days in the mountains without seeing anyone.

As I ran, I tried to ignore the boring, tan-colored box houses with quarter-mile spacing that modern people in this cultivated region called home. Through the fields to the north stood a single, narrow blue house with white trim, which from a distance looked like a cottage I had once seen on the outskirts of a muddy village on the north shore of Lake Baikal. To the southeast, across several ditch channels, was a farm with a rose-colored barn and a broad-gable roof. From afar, these looked like a farmstead I'd once visited in the Netherlands as a young professor, zealous to learn how individuals in rural communities were capturing methane in their artesian wells to use as a free source of natural gas to fuel lamps and heat water.

Could there be any such interesting resourcefulness here among my midwestern neighbors? I wasn't so sure. Television seemed a popular pastime in the rural Minnesota culture these days. My only escape to a more interesting life seemed to be these illusions and the memories of my past. But would they sustain my appetite for exploration, discovery, and beauty the remainder of my life, for what could be another sixty years wedded and bound to a single plot of Minnesota earth that was my husband's heritage? Peter once told me he had learned to see the beauty and wonder in small things. Could I learn to look down to see the beauty in small details of

plants and insects and soils and seasons, instead of searching for exotic wildernesses and diverse cultures that may not be found across
the vast midwestern seas of corn and soybean fields?

This day, like most other days for me in Minnesota, passed with
little adult interaction. After a brief exchange with Jorgen's morning babysitter, he and I returned to the farm, to the small white
house on the edge of the south field. The ranch-style house, built
in 1962, was functional for an old lady and was still furnished that
way, except for the bedroom that once had served as Peter's grandmother's television room. That room was now impassible. I'd filled
it with boxes of field and lab equipment, the supplies and tools I
used to launch a lake methane monitoring effort in Minnesota.
Wired as I was for a pattern of fieldwork, moving to Minnesota
had not stopped me. I hired one of Peter's farmworkers to help me
part-time. With Jorgen in a backpack, this helper and I deployed
and monitored automated bubble traps in Minnesota lakes. We
chainsawed ice blocks in spring to study ice-trapped bubbles, and
captured the bubble gases as I had done for many years in the Arctic. My crowded science room felt as out of place in the little white
farmhouse as I did in the state.

My favorite place in the house was the small, two-person suede
sofa covered with a protective red-and-white-striped cloth where
Jorgen and I would read books together for hours each day. When
Jorgen saw me heading in that direction, he dropped whatever he
was doing and dashed for the couch, his wild blond curls jostling
on his bobbing head. Jorgen couldn't get enough reading time,
and when we weren't reading, we were singing. Jorgen wasn't old
enough to have conversations, at least that is what I thought since I
wasn't tuned in enough to little people to know how to converse on
his terms. He toddled around the house after me, content to explore
the pots, pans, and Tupperware in the heavy wooden kitchen draw-
ers while I cut up raw chicken and vegetables for our stir-fry dinner.

Peter came into the house long after dark. Behind the squeak of the farmhouse screen door, I could hear the loud chirping of crickets in the grass outside. He took a seat in the chair by the door and started removing his boots. His blue coverall, sunburned skin, and blond hair were all a uniform color of the brown dirt. I could see that he was tired, but it was a good kind of tired. The kind of tired I used to feel after a long day of rowing around Shuchi Lake and hauling field gear and samples through the northern boreal forest. My bare feet pressed into the carpeted floor I'd traversed all day. I hadn't felt fieldwork tired in years, and I yearned for it. Sure, we did fieldwork when we were in Alaska, but I no longer pulled the long-hour stints for weeks on end the way I had before, the way Peter worked on his farm now.

My mood was dark. I knew Peter deserved to come home to a cheerful and loving wife. He was a kind, hardworking, and good husband. Harsh words rarely escaped his tongue. I understood he was working because the work needed to be done and he believed in it. He loved farming, the responsibility of stewarding his own land in the footsteps of his ancestors. As a farmer, he strived to find ways to be more productive and at the same time environmentally responsible. He believed in a family-shared, hard-work ethic and hoped that someday I'd join him, to labor at his side as he had done with me so many times before in the Arctic.

But what he got this day instead of a helpmeet was a shoulder shrug, an attempt to hide my jealousy and resentment. How could I create an air of happiness when I felt none?

During weeks of seclusion after returning from our winter-season stint in Alaska to this farmhouse, I had spiraled deep into depression. Out of a sense of urgency to keep up with my full-time responsibilities as an assistant professor, nearly every spare moment I had when Jorgen was sleeping or at his babysitter's was spent crunching data at my laptop computer, which was set up on an old, wobbly, dark wood table on the orange carpet of the living room.

Peter took off his dusty coveralls and hung them by the front door. He washed his face and hands and came into the kitchen where I had started serving his plate of supper. Minutes dragged on as Peter sat in silence, reading a farming magazine while I moved around the kitchen with my back to him, biting my lips to keep back a torrent of emotions.

Jorgen, still awake after his long afternoon nap, moved from behind my legs over to his dad, where he hopped onto Peter's large bare foot, hoping for a pony ride. Having filled Peter's plate and milk glass, I turned to set them on the table before him.

"Katey, are you okay?" Peter finally asked.

All at once a string of bitter discontentment poured off my lips.

"How long are we going to live like this? I hate this house and I hate this life!" I wailed. "You're gone all day. I feel so depressed. Can't we tear out some walls and make this place into a home that we like? Your mom spends all her free time mowing the lawn. Is that what I'm supposed to do the rest of my life when she dies? I know there's more to life out there in the world than spending thirty hours a week on a lawn mower. This is not the type of life I dreamed of living. It might be your dream, but it is not mine!"

By this time, my voice had risen to a shout.

Peter sat in silence for a few minutes, giving himself time to think. Impatient, I turned back to the counter of dirty dishes. Suddenly I jumped. Peter had slammed his fist down on the table behind me, spilling the milk glass. His face burned with anger. Frightened, Jorgen scurried to the corner of the kitchen and braced himself on the wooden cabinets. Peter had never raised his voice at me before, but now his nostrils flared and veins popped out of the sides of his neck as he stood up and slammed his chair down on the linoleum floor.

"Katey, how long are you going to wallow in self-pity? Everybody else is working hard for a common goal, for a productive, fulfilling family farm life. It is time for you to pull yourself out of this hole, become part of it, and learn to be content."

I knew in my head that Peter was right, but I didn't feel that he had compassion nor that he understood me. I wondered how he would feel if the tables were turned and he had to give up farming. It seemed that for me to become content at the farm would mean letting another part of me die, the part of me that clung to my home in the Arctic, my heart in Russia. The part of me that loved the challenge of working with colleagues, creating new scientific ideas, and turning them into reality. I was listening to the voice of pride. It told me that I could only be happy if I had things my way, if I resisted change and clung to my own former way of life.

Certainly, Peter did not know the hollowness of my deflation, the depth of my depression. The thought of facing another day of prison behind these four dark walls with nowhere to turn outside was almost more than I could bear. With despair in my heart and ugliness on my face, I answered, "If things keep on this way, you might come home to find me hanging from a grain storage bin."

We both looked at Jorgen. His little eyes stared wide and scared at us from across the kitchen. Then we looked at each other.

In silence, I acknowledged my shame.

No matter what age, no child should hear words like this from their mother. Such threats should never be uttered to a husband. Peter stormed out of the room and out the front door. I put Jorgen to bed that night alone and pleaded that God would help me. I knew he had the power and ability to free me completely from the negative attitude from which I was suffocating. I knew that pride was my problem, my lifelong attitude of wanting to have things my way. Why was I so afraid to let it go? Until I did, joy would be impossible.

"God," I prayed, "you allowed Joseph's brothers to throw him in a pit and then sell him off to slavery in a foreign land. With less choice than I had in coming to Minnesota, Joseph was carted off to Egypt because you had plans for him there. Important plans. I know you have plans for me too, otherwise I would not have trusted you when I married Peter and came here. Please, God, help me to overcome my pride and my jealousy of Peter's family's joy in living on this farm. Please help me to lay down my own will, loosen my stiff neck, and trust in you while you mold me into your likeness, a being filled with love, joy, peace, patience, kindness, goodness, gentleness, and self-control. I know that your plans never fail, and that somehow you will weave the strands of my past into the creation of something new and good, if I lay down my will and follow you."

...

THE NEXT MORNING, I awoke to my bondage of self-will only slightly loosened. But it was enough to see out through the cracks. I began to see how my life was parallel to a thermokarst lake. I had been working for months to pull together the pieces of the Siberian peat story, and now I realized that a thermokarst lake was a lot like me.

A thermokarst lake is born to promote itself at all costs. A full force of nature, it gnaws away at the icy, carbon-rich permafrost soils stored in the land around it. In its youth, the lake grows larger day by day and year by year by consuming its surroundings. It takes more than it gives, and what it does give is not always good for the world. The lake is productive and successful in its work, generating vast amounts of greenhouse gas in a selfish feedback cycle that causes more warming, more permafrost thawing, and more lake growth. Driven by self-perpetuation, the process continues blindly until the lake breeches a topographic gradient and catastrophically drains—in essence, having eaten its way to its own death. But there is an alternate fate for a self-seeking lake. Its margin waistline can become so thick with thawed soil that it can no longer effectively melt the ice around it. In this mature state of fatness, the lake lounges on the landscape, digesting, burping, and venting the dwindling resources in its gut until it runs out of food. Then its character changes. Instead of hurting the environment around it (by self-perpetuating global warming), the lake assumes a new roll. Nutrients in the lake support the growth of plants and other organisms, and as these flourish in the lake's embrace, they soak up carbon dioxide from the atmosphere and build carbon-sequestering peat on the lake bottom. This positive role of helping the environment around them by absorbing greenhouse gases and accumulating peat can continue for a very long time. The longer it continues, the better its impact on the world. By the time the old lake dies, either by yielding its life to peat infilling or by draining or drying and refreezing, the lake will have absorbed more carbon from the atmosphere than it originally belched out.[1] Its cumulative lifetime impact on the world had the potential to be positive, and so could mine, if I was willing to lay down my pride and embrace change.

Dedicating my life to scientific publications at the cost of personal relationships would not necessarily make the world a better place. Peter was supportive of my science, but he reminded me, "With the passage of time, your papers will become obsolete. It is the people in your life and their characters, which you influence, whether for better or worse, that will live on." I understood what he was saying, but some things are easier said than done, and I was addicted to my work.

In 2015, Anders was born, a baby brother to Jorgen. Bound to the farmhouse with another infant, I labored furiously on manuscripts that addressed the importance of thermokarst in our future.[2] Global climate models had begun to factor in permafrost carbon

In a thermokarst lake's life cycle, a young pond grows by consuming the surrounding permafrost soils and releasing large quantities of greenhouse gas. As the pond matures, its function changes to carbon sequestration, so that by the time it dies, its net impact on the climate was beneficial.

release, but so far they only considered upland terrestrial ecosystems, such as forests and tundra, and ignored lakes.[3] I knew that lakes should not be neglected because they occurred all across the Arctic landscape and they behaved very differently than land. Lake water retains heat throughout winter, causing permafrost to thaw far faster and deeper than on land.[4] Besides, Sergey and I had nearly two decades of data showing ancient methane ages in lake bubbles. These data pointed unwaveringly to the release of old carbon due to new permafrost thaw. Somehow I needed to prove the necessity of including thermokarst lakes in global climate models.

That is when Guido and I joined forces once again. I combined box-model output his postdoc had developed to estimate twenty-first-century permafrost carbon emissions from lakes with output from a well-accepted global climate model predicting permafrost carbon emissions from land.[5] The results were shocking! Including lakes more than doubled the climate warming impact of thawing permafrost in the twenty-first century.[6] While new lakes would occupy only a tiny fraction of the Arctic, they would be responsible for about half of all permafrost-derived warming. This meant that thermokarst lakes were hotspots of permafrost carbon emissions on the landscape. Together, permafrost thaw from lakes and land were expected to accelerate anthropogenic warming by 10 to 20 percent, making mitigation all that more difficult.[7] Of particular concern, our calculations showed that even if society succeeded in curbing fossil fuel emissions to limit global warming to 2 degrees Celsius, this warming was still enough to open the freezer door and cause widespread carbon release from lakes. The only way to prevent the large permafrost carbon feedback was to attain net negative global emissions as a society. But we were not on that trajectory. Global inventories showed atmospheric greenhouse gas concentrations were steadily climbing (and still are) the business-as-usual pathway, leading to a 3 to 5 degrees Celsius warmer Earth within the next fifty to eighty years.[8]

The seemingly dismal outlook on Earth's future was compounded by my inner resentment of being stuck in Minnesota. I knew I had much to be grateful for. Unlike my parents, who had struggled with the burden of single-parenthood, I had a wonderful husband with whom I could partner to raise our boys. We had good health and loving friends and family. The roofs over our heads, both in Alaska and in Minnesota, were debt-free and provided shelter from bad weather. Over the course of time, Peter's parents built themselves a retirement home in the place where his grandmother's house once stood, and we moved into the central farmhouse, remodeling it to our liking. We always had enough to eat, and a fair portion of our diet came from subsistence fish and berries in Alaska or meat, vegetables, and fruits we had raised ourselves in Minnesota. We had jobs we enjoyed doing. I knew that so many people on Earth were in need of such things and hence I should be grateful. But knowing in one's mind is not the same as knowing in one's heart, and my heart was still prideful. Despite all of these undeserved blessings, I was still a pitiful human being who continued to wallow in a mire of fear and selfish discontent. Our boys grew older. Peter had provided us with the opportunity for a loving home, the type I'd yearned for all of my life. But to enjoy it seemed that it would cost me giving up my desire for the freedom to wander from one vagabond adventure to another, a pattern that had started as my coping mechanism at age twelve and that later became my addiction. I continued to resent the notion of being bound to a single address on Earth, a midwestern farm that was not of my own choosing.

As the years went by, I saw that my discontentment had the power and potential to consume my husband's joy. My own was already extinguished. If my discontentment continued to go unchecked, it could rob our children of a happy home too. I had devoted much of my life and career to loving and studying nature, but what good was it to love nature and what hope was there of passing that love on to my children if I couldn't bring a selfless love to my family members? I said that I loved and trusted in God, but what kind of faith was I teaching my kids if I didn't put myself fully in his hands and let him mold me into something new?

EIGHTEEN

Spying on Methane

As the years passed, each trip to Alaska became more precious. By September 2019, Peter's parents were in their early eighties. They still worked as hard as ever, but their ability and desire to manage the farm in our absence was waning. I knew I had to make our time in Alaska count. I knew this as I sat at my laptop in my Fairbanks home office contemplating a satellite scene and that year's Alaska fieldwork.

A satellite image of Earth's surface is nothing more than a mosaic of dots. Dots swelling with information, packed so close together that, viewed from a distance, they make a picture. I paused my curser at Point Barrow, the northernmost point of land in Alaska, which divides the Beaufort Sea to the east from the Chukchi Sea to the west. This is Alaska's northern coastal plain, a region so flat that once the spongy tundra becomes saturated with water, raindrops don't know which way to flow. They pile up, creating swamps on top of permafrost that extend as far as the eye can see. As I panned toward this section of Earth in Google Earth, the main feature that came into view was lakes. Millions and millions of lakes of all sizes, shapes, and colors. Most of the lakes were oriented and elongated toward the northwest, a testament to prevalent wind direction, which churns the water into currents that preferentially erode the shores at the tips of their long axes. Even along pockets of dry land, I could see the ghost outlines of lakes that once had wandered the tundra in much the same direction. Those lakes had drained in the past, and their basins were now spotted with tiny black ponds and wetlands, which I knew would be just as much a challenge to traverse as the soggy, boggy tundra.

For several years my colleagues and I had been scheming to set foot on the ground at Sukok Lake, eighteen miles south of Utqiaġvik (a city formally known as Barrow), and we finally had a NASA grant to do it. There were no roads to Sukok Lake. Getting there meant crossing wet wilderness. I knew the traverse would be easier in winter, when the wetlands are frozen and ice on lakes is thick enough to safely cross by snow machine. Our goal was to confirm through fieldwork some particular features we were seeing in satellite images, which we suspected were methane megaseeps. A decade earlier, Peter and I had flown in airplanes to map megaseeps on a narrow north–south Alaska transect, and my student, Melanie Engram, and I had developed a method to quantify normal lake bubbling with spaceborne synthetic aperture radar.[1] But, until now no one had ever tried mapping megaseeps from satellite images over much larger landscapes. We didn't know for sure if the features we saw in the satellite images were indeed methane. But we suspected it. We suspected it enough to set aside twelve thousand dollars of our NASA research grant for a single field trip to go and look.

I stared at my screen, contemplating an image of the seven-mile long Sukok Lake sent to me by Melanie. This lake is shaped like a giant pumpkin head teetering on the shoulder of a long, narrow bottom basin. At the top of Sukok Lake, close to where the stem would be if the basin were indeed a pumpkin, a large patch of bright yellow dots was circled and labeled "Katey's Seeps."

"Katey's Seeps!" I remembered this place. How could I forget that windy, cold, −36°F late-October morning ten years ago? I had been traversing the frozen, snow-covered tundra to this lake, chasing a local rumor of megaseep methane with Nagruk, my young Iñupiat guide. I was immediately back in those memories from this lake with Nagruk.

Two hours into our snow-machine ride, my face mask had been crusted over by frozen condensation from my breath. My fingers were so numb inside my mittens, I had to trust they were still gripping the handles. It had been especially risky to be out in the bitter windchill, so far from town. I realized the change of clothes I had brought along would not suffice if we got wet. As I drove in Nagruk's tracks, I prayed for guidance. Moments later, Nagruk stopped, got off his sled, and tramped back to me in his heavy winter boots and bulky black snowsuit.

"We're about at Sukok Lake now. It's just up there," he said, pointing south. As he spoke, our eyes turned eastward to watch the tip of the sun begin to rise from the horizon. For a fraction of a second, I saw steam, like a cloud, come between us and the sun. And then, as the sun rose higher, the cloud was gone!

"Nok, did you see that? Did you see that steam? The only thing I know of that makes that kind of steam in conditions like this is open water! And if there is open water, then there is sure to be gas bubbling! No matter what happens, we have to keep our eyes fixed on that point and drive straight to it," I insisted.

My blood raced during the next fifteen minutes that it took to draw closer to the point on the horizon where we had, only for a brief moment, glimpsed the water vapor. I laughed inside my mask. "God is so good!" If we hadn't stopped just when we did, I would have had my eyes on Nagruk's trail and never would have seen that undeserved clue passed down to us from heaven.

Before us in the rose-gold light of sunrise was the immense frozen, snow-covered Sukok Lake, the far margin of which disappeared somewhere beyond the southern horizon. The lake is so huge, that if we had arrived at the lake at any other point along its shore, there was no telling how long or even if we ever would have found the black, rectangular pool of water with little round black holes dotting its edges that lay in the ice right there before us. In the daylight, we could no longer see steam ascending from the pool, but in the open water holes, streams of bubbles rose to the lake surface. It was gas! I needed to get some samples.

Nagruk had no intention of letting me walk out on thin ice to approach the open holes. The risk of getting wet was high. All too often the Iñupiat buried a person who unexpectedly fell through thin ice in the wilderness, and Nagruk's job was to bring me back to Utqiaġvik safely.

I was able to convince Nagruk, more than myself, that the nylon Alpacka raft on my sled would keep me afloat should I start to break through the ice. But secretly I doubted the integrity of this synthetic boat. I'd seen before how plastic becomes brittle and shatters in the cold. It was quite possible that a combination of sharp ice and frigid temperatures would be enough to puncture my raft. But I didn't let Nagruk know I was thinking this way. This was an opportunity I could not pass up. With my left foot rooted in the inflated boat and an ice spear in my right hand, I moved forward, dragging the boat out onto the ice, one cautious step at a time. In a matter of minutes, I was at the edge of a bubbling ice hole. I reached for a trap and collected gas.

Months later these samples would reveal that the bubbles were from the same reservoir of natural gas hundreds of meters belowground, which the residents of Utqiaġvik tapped into for energy via human-made gas wells. But there at the tundra lake, the gas was leaking naturally on its own into the atmosphere.

Ten years later, on this September morning in 2019, I reviewed Melanie's PowerPoint presentation. My attention was drawn to a spot three miles south of "Katey's Seeps," a place labeled "Guido's Line." In a satellite radar image, Guido's Line was a streak of yellow dots occurring in exactly the same position in the lake's southern basin as where Melanie and I had seen dark, open-water circles aligning in optical satellite imagery Guido had shared with us. In addition to Guido's Line, several other potential seep locations were labeled "New Seeps."

New seeps! If Melanie was right, if these yellow dots did indeed indicate gas seeps that had not been there in 1992—the year the earliest available radar images were taken—then this would be evidence that permafrost in the Arctic was actively changing, not just at the surface but at depths far belowground too. Opening this permafrost seal would mean more methane escaping into the atmosphere than anyone anticipated. The natural release of this fossil fuel could lead to only one thing: an acceleration of climate warming.

Verifying that we were indeed able to detect methane megaseeps with satellite images and then to explore the hypothesis that new seeps were appearing in lakes where they had not previously been detected in the earliest available satellite radar imagery was my mission for 2019's fieldwork in Alaska.

...

SUDDENLY THERE WAS a dull thump on the door of my downstairs office. It opened, and Jorgen, still in his pajamas and rubbing his eyes, came over and sat on my lap. He was tall and strong for a seven-year-old, but the person of few words inside this sturdy frame responded best to gentleness. His curly head tickled my cheek as I squeezed him from behind.

"Daddy says breakfast is ready," Jorgen reported. We hopped off the swivel chair and raced up through the dark, wood-paneled hallway to emerge in the sun-drenched main room of our two-story cabin-like home in Fairbanks. Anders was already seated in his high chair at the table by the window, showing his stuffed-animal lamb a

LEGO trailer he had built. In front of him, yellow pillars of orange-juice glasses gleamed in the sun's rays. Tendrils of steam wafted up from four plates of toast and poached eggs. Peter was just scooping vanilla ice cream onto bowls of oatmeal, a novelty I'd gotten used to after moving to his farm. Jorgen and I took our places at the table.

While we were still eating, Peter read aloud a chapter from the Bible and then pulled out the boys' homeschool memory work. Enthralled with maps, Jorgen already had several sheets of paper in front of him. On one he was labeling European water bodies and on another, in between bites of oatmeal, he was filling in details of his rendition of the land of Narnia. Anders, a four-year-old who still preferred to be spoon-fed his eggs by a parent, was as loud as anybody else in singing multiplication tables and chanting Latin conjugations, meanwhile dancing his lamb on the table beside him. The energy level at the table rose. Soon Jorgen's wiggling shook the whole table and Anders giggled so loudly that we knew we better send them outside to play before starting math.

I prepared two tall lattes for Peter and myself and then went out to sit beside my tall, handsome husband on the deck. It was good to be with him here in Fairbanks. There was no grain auger he needed to run out to shut off or truckload of soybeans to haul to market. Here Peter wore the same Carhartt pants, T-shirt, and navy-blue sweatshirt he wore on the Minnesota farm, but far from the daily concerns of farm management, he would sit down, cross his legs, and simply enjoy these moments with his family. Behind him a hundred golden birch leaves fluttered to the ground, reminding us that our days of feeling the sun's warmth on our bare skin in Alaska were numbered.

"Winter is on its way," the breeze whispered through our yard, causing the hairs on my arms to stand with the goose bumps. The magenta flowers on the seven-foot-tall fireweed stand below had long senesced, and in their place silky strands of seeds swirled out of burst pods. From somewhere inside the fireweed stand, Jorgen

and Anders emerged, with wooden survey-stake swords in one hand and metal garbage-can-lid shields in the other. Roars of attack were sounded as the boys dashed to the woodshed. Peter built the shed following years of our work together felling trees to clear a portal on our Chena Ridge property through the forest out onto the Tanana River and its floodplain. The boys clambered up six-foot piles of split birch and poplar to traverse the woodshed's beams and rafters ten feet above the ground, their weapons tucked under their arms. They had been climbing the shed before its construction was ever complete, but still I could not bring myself to watch them, especially little Anders, whose bottom stuck out so far that he looked unbalanced as his little feet maneuvered from one rafter to another.

"Peter, I've been thinking that my students and I should go to Utqiaġvik in November. Frozen ground conditions for getting to Sukok Lake should be good then, and seeps are easier to spot in winter since they make ice holes."

Peter sipped his coffee. With a slightly furrowed brow, he gazed off past the woodshed toward the northwest horizon, where the top of Chena Ridge was laced with paper birch and white spruce trees. Through ten years of marriage, I had improved at yielding to patience. I knew his well-thought-out response was better than a fast one.

"Now is the time to go," Peter finally said. "Now, when the days are still long and the sun is somewhat strong."

Ten years ago I would have jumped at the spontaneous opportunity for a remote and wild field trip, but not now. I did not want to go on a field trip alone, leaving my family behind. In nine days we needed to return to Minnesota for fall crop harvest. In Minnesota, there would be long months when the boys and I didn't see much of Peter while he was combining grain. For our last nine days in Alaska, we had plans as a family to wrap up fieldwork on local lakes, pick cranberries, and hike. My campus calendar was full of meetings too. Besides, the logistics of remote field trips take weeks to plan.

I complained to Peter, "I've been working through weekends all the past month. I was hoping to finish a data set this weekend so that I could hike in the fall colors with you and the boys next Saturday."

"We are not in Alaska for leisure, Katey. If there is important science to be done, you must go do it," Peter insisted.

"You know how hard it is to find gas seeps when there is no ice on lakes," I argued. "I would need a perfectly calm day, a day when the lake's surface is as still as glass. What are the chances of calm conditions?"

I went down to my home office and placed a call to the Ukpeaġvik Iñupiat Corporation (UIC) in Utqiaġvik. On the other end of the line, the logistics coordinator reported that no, there were no floatplanes in Utqiaġvik, but yes, it would be possible to reach Su-kok Lake using four-wheelers across the tundra. But the weather didn't look good. The wind had been blowing there for weeks, and the forecast called for more of the same.

I was not enthusiastic about this trip. I would have to operate a motorboat on my own, and the forecast of 33°F, twenty-mile-per-hour winds, and rain would not be prime conditions for finding methane bubbles. Besides, what a miserable way to spend my time away from my family! I went back upstairs and told Peter that I thought we should take it as a sign from God that if the wind died down, I should go to Utqiaġvik. We left it at that.

A good farmer watches the weather. And Peter did. The forecast for Utqiaġvik from Saturday morning through Wednesday was persistent wind, fifteen to twenty miles per hour, gusts to forty miles per hour. Meanwhile in Fairbanks, it was another crisp, sunny morning. I read Bible stories on the couch. Anders, still in his pajamas, warmed my lap, and Jorgen snuggled in close beside. Across the room, Peter stirred the oatmeal while paging through a farm magazine. Following breakfast, Peter and Jorgen worked on adding fractions at the table while I kneeled on the floor in front of

Anders with a cello bow in hand. Anders also had a cello bow and was seated on a stepstool with his violin-size cello resting in front of him and a penny balancing on each stockinged foot to help keep them from wriggling. He lifted his bow upright. I leaned forward, straining my neck to see behind his chubby little hand. Was his thumb bent, like a king sitting on his throne? It was! Pleased, Anders smiled as together we raised our bows in the air and sang:

Up like a rocket, down like the rain,
put your bow where you keep your brain.

Landing the bottom of his bow on top of his skull, Anders added his signature "Ow!" to the end of the song and giggled. A penny lay on the floor beside his left foot.

"We should pick cranberries today," Peter suggested as Jorgen gathered his cello and carried it outside for his turn to practice.

"Yes! It's a beautiful day for it!" I cheered, following Jorgen with a couple of stocking caps and my own cello. Jorgen took a seat, pulled his cello upright against his tall torso, and adjusted his feet, which had learned by now to remain perfectly still. The sun hit his cello, revealing long streaks that ran down the golden wood of its upper bout. Those were the tracks of his tears, the dried-up marks of the pain and frustration he felt on days when sleep had been too short or Mama too critical. But he had never given up. The worst threat I could give was to suggest that we lay down our cellos during a practice, stop trying, and walk away. "Nooooo," he would wail when I threatened to put my cello away. Jorgen was always determined to work through the challenge, but he needed two things I lacked: patience and gentleness.

On many days there were no tears, and this was one of those days. We practiced drills his teacher had assigned and then launched into the hard parts, the fifth and seventh position shifts of our newest piece, Antonín Dvořák's *Humoresque*.

"Play that part again," I insisted. "You've got to lift your elbow before shifting. Imagine the female archer in this song lifting her elbow, taking aim before she shoots. It's like that. When your fingers shoot down the fingerboard in the shift, you're in position to land on target."

The goal, according to the Japanese philosopher and music educator we followed, is to practice the hard parts, not until you get them right, but until you cannot get them wrong. We weren't there yet, but after a few practice shifts, I allowed Jorgen to do what he'd been aching to do ever since he sat down: simply play. He placed his bow on the D string and let it rip into song. His fingers pranced on the fingerboard as his bow dashed through the short, jovial notes of *Humoresque*'s jester section. I sat back and listened. His style followed the names we'd given to different sections of the piece. With a long, drawn-out F natural slurring into an E, he transitioned into the Princess section, legato, on the high A, and I could imagine Her Majesty speaking from her high throne. Jorgen dropped his shoulder to make way for his bow on the G string. Low, strong tones of the Knight section reverberated from his cello. I could feel the music resonate through the wood of our deck into my feet. The Jester and Princess each spoke in turn again, and when the piece ended, Jorgen stood up and bowed, and I clapped, as we were trained to do through the Suzuki method ever since he was four years old.

"Mama," he asked. "Can I stay out here to play a little longer?"

"Oh, please do!" I said, and I went into the house to pack a picnic lunch for our afternoon outing to a secret berry-picking site, east of home on the bluffs overlooking the Tanana River.

I could still hear the music of Jorgen's cello as I made my way downstairs. I knew the rest of the morning Peter would read to the boys, and then they'd do math. I would sit behind my laptop, crunching lake-bubble-chemistry data, and trying not to think about the possibility of an impending field trip to Utqiaġvik.

...

SUNDAY MORNING CAME, and Peter's weather forecast report for Utqiaġvik was the same. Steady wind and rain. Selfishly I was glad for the bad weather forecast, so glad that I started to relax in my thinking and make other non-travel plans for the week.

Monday morning, as usual, I rose in the dark while everyone else was asleep. As the teakettle heated up, I read the Bible. Then, with hot tea in hand, I tiptoed down to my computer, where I concertedly avoided looking at email in order to write a manuscript without interruption for a couple of hours. Eventually I heard the clamber of Anders climbing out of the wooden crib where he still slept in our bedroom, then the patter of his bare feet as he scurried to the bathroom in the morning twilight. The rest of the house was quiet, so I knew Jorgen must still be asleep on his bed, a mattress on the floor of our bedroom closet.

Soon, Peter appeared at my door. Anders was in his arms, burying his face in his dad's shoulder to get away from the bright incandescent light in my office. "Anders is wondering if you will come upstairs and read to him," Peter said. Somewhat reluctant to leave my train of thought at the keyboard, I agreed, knowing it was the right thing to do.

Moments later, snuggled up in the soft lamplight with Anders, his little stuffed lamb, and a tall stack of books on the couch, the thought train that had been chugging across my computer was out of mind. My son was here now on my lap, pressing into me for warmth, soaking up the words from the page and taking them into his mind wherever the millions of brain signals were taking them as he grew, more slowly than I could measure. But I knew from morning after morning of this same ritual that he was growing. He wasn't any longer the tiny infant who balanced between my breast and arm, unable to hold up his own head. Nor was he the toddler who couldn't stand more than a few minutes of picture books (unless they were about tractors) before wriggling away. He was a little boy now, almost too heavy to lift off the ground. And he loved to listen to stories.

On the way to the breakfast table, I asked Peter what the forecast said.

"It calls for flat water Wednesday morning at 9 a.m.," he answered in a deep, quiet monotone that sometimes made me wonder if his Scandinavian blood caused him to overly dampen emotions at times when mine would surge. "Winds slowing on Tuesday. Wednesday morning will be dead calm with winds picking up after noon Wednesday and Thursday."

I almost dropped my water glass. "What? The forecast changed, just overnight?"

"That is what it says," Peter answered.

Moments later, my cell phone rang. UIC was on the line from Utqiaġvik. "It looks like the conditions you were asking for will be here tomorrow."

Tomorrow! Tomorrow was booked with cello lessons, a dentist appointment, and a long string of meetings at the university. I could hardly swallow my food. I remembered what I had said about a change in weather being a sign from God, but I didn't bring it up. To reach Sukok Lake by tomorrow night I would have to leave that night or possibly the next morning. With all that still needed to be done, that night would be impossible. I helped clear the table and went downstairs to look at flights.

"Wilbur," I said to the man at UIC, "the soonest I could get to Utqiaġvik is on the flight arriving at 6:30 p.m. tomorrow evening. Do you think it is remotely possible to get out to Sukok Lake that night, or will that be too late to start the three-hour trek by ATV? I must reach the lake's shore by Tuesday night; my window for sampling is first thing Wednesday morning.

"No, it won't be too late," Wilbur replied. "I will ask Nagruk if you can use snow machines instead of ATVs."

"Snow machines? But there is no snow!" I exclaimed into the phone.

"It's okay. You can use them," Wilbur said. "It will be faster than an ATV."

I trusted he knew what he was talking about and told him they could prepare for my arrival. Wilbur was going to ask Nagruk; it gladdened my heart to hear that the responsible young guide I'd had ten years ago at Sukok Lake was now in a position of authority.

I went upstairs just as Peter finished math with the boys and was getting ready to send them outside for a recess. I followed him out, and we took our adjacent seats on the wooden deck bench.

"I wish you could go with me, Peter," I said, looking into his face. We both knew that we were a team that could get just about any job done, if there was a job that could be done.

Peter's eyes lit up. "Why didn't you ask before?"

I hadn't asked because we'd never left our children without a parent. It hadn't even seemed like an option. Going with Peter alone would be like the good old days, when science was romantic. Suddenly the thought of going to Utqiaġvik didn't seem so gloomy and foreboding. For the first time, I felt a tinge of excitement.

I phoned the girl who rented the downstairs mother-in-law apartment in our house. She was a single, twenty-six-year-old cross-country ski coach, the oldest of six kids raised in a Pennsylvania homeschool family, a fiddler, an avid wilderness explorer, and she was fun. We told her if she stayed with our boys, we'd pay her better than the sub job she had. She agreed. If all went well, we'd be home by midnight Wednesday.

Peter came down to my office. "Is this wise?" he asked. "What if something happens to both of us out there and neither of us comes back?" We knew there were risks involved in this type of fieldwork, and neither of us wanted to do something foolish that would orphan our children.

"I suppose we could die in a car crash on Chena Ridge Road if that were part of God's plan," I said. Then I reminded him about taking the weather as a green light from God to go, and he agreed.

...

WHEN WE REACHED Utqiaġvik, a tall, young Iñupiat man with a round, gentle face appeared at the baggage claim. "Are you Katey?"

"Yes, I am," I said.

"It is nice to meet you! I'm Wilbur. I will be your guide to Sukok Lake tonight. Once you have your things, we can load them into the truck. I can take you down the road to where the snow machines are waiting."

Peter retrieved our newly purchased boat motor at cargo while I waited in Wilbur's back seat. "Aren't you worried about it getting dark?" I asked him.

"No, not too worried," came the flat answer from the driver's seat. He didn't turn around.

"Have you ever driven a snow machine in the summer without snow before?" I continued, still in curious disbelief that this was a good idea.

"Yeah, but it was east of Utqiaġvik, not down where we are headed. We are going to have to cross some rivers and ponds."

Then Wilbur turned around to look at me. A huge smile suddenly spread across his face. "My cousins and I did that when we were boys. You have to go really fast. It is sort of like a hydroplane. Just wear a good rainsuit. You get wet."

He was quiet for a moment before adding, "The rivers we need to cross tonight are deep, but we can find some narrow places."

Nervous, I asked Wilbur if anything could go wrong.

"We could get bogged down."

Peter climbed into the truck and we drove out of town.

...

A SNOWY OWL glided across the red tundra and stopped a hundred meters from where two snow machines and sleds were standing next to the road. Wearing our insulated Xtratuf boots, Peter and I pulled out the remainder of the gear we'd need to stay warm and dry on the three-hour snow-machine drive: down coats, Gore-Tex rainsuits, face masks, hats, and neoprene gloves. The plan was for Wilbur to leave us at the lake for the night, then return the next day. We decided to turn on the GPS unit in Peter's pocket just in case we needed to track our route back alone. We loaded our dry bags, equipment cases, and backpacks, along with the boat, motor, and fuel can, onto the two Siglin sleds. Having worked with Arctic residents before, we knew it was best to let Wilbur wrap the gear in tarps and lash the load down with the sleds' side ropes.

I took my seat behind Peter on one snow machine and pulled back the thick layers of coats to look at my wristwatch. It was 8:30 p.m. We'd left the boys eight hours ago. They'd be in bed by now.

The sun had set, but the lingering twilight allowed us to see the surrounding tundra. "You'll have to drive fast," Wilbur said to Peter, "especially over the wet areas."

The machines fired up and we were off, traveling much faster than I had expected. I gripped the passenger handles for balance and peered around Peter's shoulder as Wilbur sped off in front of us. Arcs of cold, reddish water shot out behind his sled as he glided across thermokarst puddles, which were ground-ice melt holes of

Wet sedge tundra pond

various depths filled with water. The spray instantly coated the load on Wilbur's sled and sloshed around the gasoline fuel can. Water and mud spewed up from the track and skis of our machine too. The outsides of my Gore-Tex pants and neoprene gloves were drenched. With a tight grip on my handles, I turned around to check on our own load, then buried my face in Peter's back to keep dry.

It was a struggle to keep up with Wilbur. The weight of an extra person on our machine did not help, especially when the ground became bumpy. Our machine lurched this way and that, splashing through thermokarst puddles formed between five-foot-tall mounds of permafrost soil. I focused all of my attention on keeping my center of gravity on the seat and my feet planted on the running boards.

When the path smoothed once again, I opened my eyes. We were speeding across wide-open wilderness, in the middle of a large, flat drained lake basin. Black silhouettes of grass and sedge stems emerged through the shallow water, appearing in the headlights of our snow machine, like a patchwork of rice paddy fields.

I could make out the dark form of a river that snaked along eastward beside us. *This must be the river Wilbur mentioned*, I thought. I watched as it narrowed and widened. If we were going to hydroplane across this river, we'd need gently sloping margins. At the speed we were going, it could be disastrous to hit a steep bluff on the far side, but it was hard to see in the dark. Finally, Wilbur stopped and waited for us to pull up beside him.

"The river is deep, but it is not too wide. Are you comfortable crossing?" he asked.

"Yes," Peter answered.

I wasn't comfortable but didn't think we had any choice.

"Just follow my path exactly, and go as fast as you can," Wilbur advised.

We watched as he made a wide curve, orienting himself perpendicular to the river. Then he accelerated down the bluff, dashed across the river, and ran up the smooth, far bank. Peter did not hesitate to follow. At full throttle, our machine with sled in tow whipped across the black snake's back.

The other side of the river was wet too, and our machine slowed down in the long stretches of water. The cone of Wilbur's headlight shrunk as he gained distance from us. There were no longer any mossy places for our machine's track to grip into, just flooded wetlands as far as we could see. Peter gave the machine full gas. Our engine whirred, and water flew in one long continuous spray. At full throttle we were still losing speed. The smell of burning rubber filled my nostrils. We'd slowed so much that clouds of vapor or smoke—I didn't know which—surrounded my head. Peter stood up, attempting to lurch the machine forward with his own body before we finally came to a full stop.

I jumped off in alarm and shouted, "Is our machine going to catch on fire?"

"No," Peter commanded sternly. "Get back on."

I looked down at my feet. The water was rising slowly up my calves as I felt myself sinking. I realized that we were stuck in a floating bog. I looked over at our machine, which was submerged up to the hull in the same mire. I jumped back onto the running board.

"What a form of transportation this is!" Peter said in disgust. "The clutch is full of water so the drive belt cannot engage. What are we going to do now?"

In the distance, Wilbur's headlights were a small, ever-shrinking point on the horizon. "He can't stop here," I wailed, as a feeling of panic started to overcome me. I turned around, searching in vain for a place in the swamp to make a shelter. I was deeply glad that we hadn't brought Jorgen and Anders along with us, as we had on so many field trips in the past.

"Look over there." Peter pointed to the west. Against the clear midnight-blue sky we could see the black silhouettes of a dozen or more caribou antlers bobbing along. They weren't running.

"At least we know they are not being chased by a bear nearby," Peter comforted. Fortunately, the headlights of Wilbur's machine grew larger again. He came to a stop on a patch of moss about thirty feet from us and waded over. He and Peter worked together to nudge the skis up out of the peat and tip the machine to one side, letting water drain out of the clutch. Wilbur pointed southeast. "Let's turn and try going that way," he said. "Maybe there won't be so much water there."

"Thank God," I said as we turned and finally drew up next to Wilbur on the only mossy place large enough to hold two machines.

"We better stop here," Wilbur suggested. Peter took out the GPS unit. We were on the northeast margin of Sukok Lake's giant, pumpkin-head-shaped upper basin. "We'll have about a three-mile boat ride across the lake to reach Guido's Line in the morning," Peter judged.

"I'll be back for you tomorrow at 1:30 p.m.," Wilbur said and then sped off into the darkness.

Peter found a soft, mossy mound, an inch or two more elevated than where we stood, to pitch our tent. We didn't mind that it was covered in goose poop; at least we'd be drier there. I undressed as quickly as I could to climb into my sleeping bag. For the first time, in the beam of my headlamp after sitting down, I realized that the dark wet stuff coating me and everything on our sleds wasn't mud but tiny moss leaves. I remembered hours spent in the Northeast Science Station in Siberia with Peter, peering through a microscope and picking with a pair of forceps tiny moss leaves just like these from our alas samples. This night, we'd been churning the tundra vegetation as we drove and were now completely covered in moss-leaf soup.

I went to sleep thinking of all Peter went through to help me in science. Would I ever help him that way on the farm? Would I ever share in his toils and be his true helper? I knew that was his heart's desire. I tried to think of what I could possibly do to be more of a farmer myself, but nothing came to mind. I didn't have mechanical skills, and long hours of truck and tractor driving still didn't seem appealing. The call of an unfamiliar bird carried across the lake, and I fell asleep.

...

"THERE IS MORE light in the sky now than when we went to bed," Peter said, rousing me at dawn a few hours later. The lake's calm surface was visible through the back door of our tent. I dressed quickly, pulling the heavy, wet Gore-Tex suit on over my dry layers of SmartWool long underwear and clothes. Peter was inflating the boat by the time I crawled out of the tent. Low clouds hung in the sky across the vast tundra. The sky here was even bigger than in the Midwest. Maybe Minnesota wasn't so flat after all.

By the time we loaded our boat, a slight breeze had picked up across the lake. I worked the oars to keep our raft from blowing to shore while Peter pulled on the motor's starter rope one, two, three times. Nothing happened. He pulled hard again, this time five or six times. Still nothing. Sweat started to bead on his forehead. He ripped off his hat, life vest, parka, down jacket, and wool sweater—everything down to a single layer of long underwear. I tossed an anchor to keep us offshore. I didn't know how long it would take to get the motor started. We both knew it was dangerous to get wet in the Arctic, and sweat was a form of wet. Peter pulled again on the starter rope but to no avail. Reexamining the choke and connections to the fuel can, we wondered if water had somehow gotten into the fuel on the snow-machine trip out.

Rowing to Guido's Line was not an option. I had plenty of experience rowing across large lakes before. You can pull on the oars for hours not getting very far even with the slightest breeze. Besides, the wind was already starting to pick up. I knew God had not let us come this far only to have a failed motor keep us from reaching Guido's Line.

As Peter swiveled the motor, he noticed in the far back the gas valve. "Aha!" he exclaimed, and hope did a somersault in my heart. The motor spun right up. I quickly hauled up the anchor, and we were off at the speed of 2.4 horses pulling us through chest-high water.

I placed a polyethylene bubble trap over my lap to shield the splashing water. In no time at all, the black, peaty lake margin where we had camped became an optical illusion: a beaded necklace with dark pearls of peat pinned between the horizon of silver water and gray clouds. Then, we could see no land at all, only water meeting the sky in all directions. "This is more like a sea than a lake," Peter remarked. He kept the raft pointed toward the southwest and eventually slowed down. I glanced at the GPS unit. We were approaching Guido's Line.

I knew what to do. Refocusing my vision to detect bubbles was, after all these years, second nature. It was what I was drawn to do. Whenever I passed a calm body of water anywhere in the world, whether in a vehicle or on foot, I lost all sense of thought and conversation and stared at the water, searching for bubbles. This day, the surface of Sukok Lake was wavy, making bubble scouting a challenge. Small swells of water rose and fell. Contours of the swells had dark lines and shadows, and streaks of tiny bubbles formed as small waves crested and broke. Fortunately, ebullition bubbles radiate outward in a circular pattern, distinct from the linear crest of breaking waves. Peering into the water column, I could see the gray-greenness of depth but no rising bubbles.

The motor diminished to a quiet putter. Peter navigated slowly, referring to the GPS screen. "We're at the southeastern coordinates for Guido's Line," he said. For several long minutes we scanned the water. There was no sign of bubbles. I wondered how wide Guido's Line was and whether we should drive a zigzag pattern across it. Suddenly Peter called out, "There!" I looked and immediately saw what had drawn his attention. A circular plume of bubbles was breaking the surface of the water, over and over again in one spot. We drove in that direction and were soon surrounded by tens of bubble plumes, easily visible, rising through the lake's water column like upward-floating streams of semitransparent bath oil beads.

Then began the game of chase. Orienting the boat to hold a submerged bubble trap steady over a gas plume was not easy. The wind was calm, compared to an average day on Alaska's North Slope, but even the light breeze was enough to blow us and our boat off course. Finally, I aimed for a bubble plume and dropped a trap below the water's surface. My aim was accurate. In a matter of minutes, the clear plastic bottle from the sunken trap bobbed back up to the lake's surface, full of gas. Success!

"Take lots of replicate samples," Peter advised as I transferred gas from the trap into bottles. "It took a lot to get here. We might as well do a thorough job."

We moved farther along Guido's Line toward the western lakeshore. Bubble plumes became more numerous, more obvious, and seemingly larger as we entered the shore's wind shadow, where the water's surface was smooth. We tracked the line of bubble plumes with our GPS unit, then moved off Guido's Line. Off the line, the bubbling plumes disappeared. This meant that Guido and Melanie had mapped out the megaseeps perfectly from satellite images! Exhilarated, I couldn't wait to tell them the news.

My watch read 10:30 a.m. I wanted to see if we'd find bubbles at the spots marked as "New Seeps." As we drove toward our next set of coordinates, located nearer the lake center, the wind picked up and generated bigger waves that sent white streaks of froth streaming across the lake. Arriving at our destination, Peter killed the engine, and I stood up, hoping for a better view. I saw nothing but the thousands of tiny bubbles forming by breaking waves. It was impossible to discern methane bubbles if they were there. I was cold and wet. But Peter was in no hurry to give up. He turned the motor on, adjusted our location, and used every minute we had available until it was time to get back to meet Wilbur. As we drove away from the site, not knowing whether or not the "New Seeps" had bubbling methane, I was disappointed. How wonderful it would have been to confirm bubbling at all the locations on Sukok Lake. Peter was thrilled for the success of one site. I was grateful for that too, but we left with more questions than answers. In the days that followed, we would celebrate with Melanie and Guido the initial success of having proved with fieldwork that our interpretation of the satellite imagery had been correct. "We should apply for a patent," Guido suggested. "Remote sensing will be a powerful new tool for mapping natural gas megaseep emissions. That is still one of the wild cards in climate change."

The snow-machine trip back to Utqiaġvik from Sukok Lake was more pleasant in the daylight. The dark rice-paddy-type pools we'd crossed at night reflected the afternoon's blue sky. Crimson-red stems of *Arctophila fulva*, commonly known as pendant grass, emerged from the water where the ponds were deep. In the shallow areas, senescing sedges were a rainbow of colors: green faded into yellow, orange, and red moving up the leaves toward their gray-white tips, off of which cool September sunlight bounced. Families of ducks took to flight as we streamed past their summer home.

We traversed the wilderness, moving peacefully through its beauty. Part of me longed to stay there, to take up residence in the lakeside hunting cabin we passed, where the tundra became hilly again. But we are not designed to remain in the moments of inspiration. The test of our spiritual life is the power to descend back into the valley of ordinary stuff, where God has work for us to do.

Ten hours later, Peter and I quietly opened the front door of our Chena Ridge home and tiptoed upstairs to our bedroom. The hallway light illuminated Anders's round cheeks on his pillow. His little lamb rested on his outstretched palm, and behind the closet door, Jorgen sighed in his sleep.

NINETEEN

Love at Last

And now abide faith, hope, love, these three;
but the greatest of these *is* love.

1 CORINTHIANS 13:13, NKJV

I n early November 2019, I looked out through our farm kitchen window. The fields surrounding our house were a flat yellow. I could no longer see the rich, black Minnesota prairie soil. It was covered by sharp eight-inch spikes of dry corn stalks projecting from a carpet of loose husks, shredded leaves, and cob fragments—all that had shot out behind the combine, all that was not grain.

I yearned to get out of the house and run, but I had to stay indoors. Peter would be combining all day on fields farther from our house, and it was my job in Minnesota to get the boys through their homeschool lessons. After that I had teleconferences and a long list of deadlines for the university as well as my own self-imposed academic projects. I could get started on none of this until Anders finished his eggs, but like all mornings, he was taking his own sweet

time. He was thinking up jokes and laughing with Jorgen more often than he was putting bites into his mouth. Jorgen had his math
books spread out on the table in front of him, but his attention was
drawn to Anders. Each time I glared at Jorgen to get back to work,
his smile faded and his sullen eyes fell to his worksheets. Tension
mounted inside me. *If I don't tighten my control on these boys, we
won't get anything done today*, I thought. I'd forgotten that the primary reason we'd decided to homeschool was that we took pleasure
in our children. We enjoyed who they were, learning alongside and
building relationships with them. The root of the word "enjoy" is
"joy." But my joy was often far from complete in Minnesota, and
on this day, it was particularly lacking. I shouted from the kitchen
sink, "Hurry up and eat, Anders. Jorgen, stop distracting Anders!
Go upstairs and get to work on your math!"

Anders continued joking. I turned around to give him a stern
look and noticed the gaping hole in his top front teeth as he
smiled. The hole was there from the time a year prior when I'd
tried to force-feed him a cherry tomato. In protest he had leaned
back and fallen from his stool at the kitchen island with a fork in
his mouth. His entire three-year-old front tooth and all of its long
root ejected from his bloody mouth and lay on the floor beside
him. But at this moment, Anders wasn't thinking about his tooth.
He was lost in a world of funny thoughts and ideas. Disregarding
me completely, he slipped down off his stool and pattered out of
the room.

"Where are you going?" I shrieked with my hands still sunk in a
basin of dishes. "Come back and finish your breakfast!"

He disappeared into the bathroom. Ten minutes later, he was
still in there. I opened the door and saw him examining the faucet's
laminar flow of water over his hands.

"Stop playing in the water! Come finish your breakfast!" I
howled.

"I need to get something!" Anders shouted back at me and darted for the playroom.

"No, come now to the table!" My sudsy, wet hands were on my hips. All I could think about was the mounting pressure of what I needed to do this day and how things would go so much better if this four-year-old would work as efficiently as I did.

I could hear Peter, in the adjacent room, get up from his desk and head for the back door. "Don't forget to take your lunch. Your pail is on the back step," I blared across the house.

My head was aching. The sound of Anders's loud chatter in the kitchen began to pierce my eardrums, so I fetched earplugs to dampen the sound. Halfway through the morning, I found myself going hoarse from my own incessant commands.

Peter was home that night for supper. The sun had set, and the boys were fast asleep in their beds. Peter and I had made plans days ago for some time together that evening, but in my weariness and stress, I'd forgotten those plans. The notion that strong marriages are built on a foundation of prioritizing the relationship was far from my mind. Suddenly remembering our plans, I said, "I'll be there in fifteen minutes" and then went back to my computer to finish up some pressing work-related emails. Silently Peter went to work on his own computer. When I was done, I traipsed upstairs, assuming he would follow me, and started to get ready for bed. At the front of my mind was the fact that I needed to get up early in the morning, to get important work done before the boys woke up. Peter remained downstairs at his desk. I brushed my teeth and then whispered harshly from the top of the stairs down to Peter, "I am going to sleep in three minutes. Are you coming up or not?"

I went back to the bathroom to finish getting ready for bed. Moments later, Peter appeared at the door. He raised his fist into the air and slammed it down into his hand.

"I've been waiting for you for the past fifteen minutes while you were emailing!" he hissed. "Now I need to go fuel the combine. I'll be working until four in the morning. I have a lot to get done to finish harvesting before snow comes Thursday."

I was shocked. Peter never spoke so harshly to me. "That wasn't nice," I accused. "Why are you talking to me like that?"

"Because it's how you talk to everyone else. You shouted at the kids all day, and you've been commanding me all night."

His words stung my ears. They cut to my heart because the truth hurts.

"Okay," I said. I was utterly ashamed. "You better go."

Peter left.

I sat in the bathroom, steeping in remorse. He was right. I'd spent the morning with earplugs in my own ears all through home-school to dampen the sound of Anders's shrill voice, but really it was my own that I couldn't stand. I shut off the bathroom light and walked down the hallway in darkness past the bedrooms of our sleeping children.

To my great surprise, Peter was in our bed with his knees folded up under the white down comforter, which cascaded over his body like a snow-covered mountain. Lamplight reflected warmly from his clean-shaven face.

"Y-you're still here?" I stuttered. "I . . . I thought you were outside fueling the combine. If I'd known you were here, I wouldn't have stayed so long in the bathroom."

I sat down on the bed, half undressed, in pursuit of pajamas but still wearing my black wool shirt from the day. My shoulders slumped, and I turned away from Peter. I couldn't bring myself to sit tall. In my disgrace, I could not face him.

He continued in a soft way to berate me. "It is overwhelming for little Anders to absorb your incessant commands all day long. This morning you harped on him every two minutes to eat his breakfast. A soft suggestion every fifteen minutes would probably be more effective. You need to be more judicious with your words. Speak less often and I think you'll find that your words will have more impact."

I knew he was right. I also appreciated the suggestion for how to turn the corner to do what was right. That was better than leaving me with the accusation of my trespasses. Against my own will, I brought myself to look at him. There was a question on my mind. Why was it he was lying undressed beneath the covers leading this painful conversation about my shortcomings? It seemed so out of place. Probing deeply into my wounds seemed like a strange way to set the stage for marital intimacy. Yet his gentle pulling away at the outer layers of my being and prodding into the most tender and vulnerable center of my weaknesses was more intimate than physical touch.

"I don't want to be that way, so controlling," I confessed as the pride in my heart started to dissolve. Tears began to stream down my cheeks, each drop carrying away with it some of my terseness. "As it is, we don't make it through all the homeschool material that we should in a day. If I loosen the reins, I'm afraid we won't get anything done. I just want the boys to do well and be well," I explained, sobbing.

"We know you do," Peter answered. There was a loving chuckle in his voice. Then he was quiet. After thinking, he continued. "It is just that ordering people to do well and be well on your terms doesn't always bring out the joy in life."

I thought about what he was saying and about how it had never been very fun to play as a child when I spent most of my energy trying to control the other kids.

"I want a gentle wife and gentle mother for the boys," Peter continued. He'd rarely ever stated anything in terms of what he wanted, but now he felt it was necessary to clarify that it wasn't a different wife that he wanted but that he wanted me to be gentle.

"I will try," I said. But I knew that gentle was something that

I was not. I found I didn't really even know what "gentle" meant. I had a book about gentleness that I had yet to take to heart, and I recalled that in it, gentleness was defined as "kind, free from harshness, willingness to yield, soft in action."[1] Most of the time, I was so dead set on pursuing my goals, bulldozing anyone and anything that got in my way, that I was not any of those things. The book said that becoming gentle would "not imply becoming weak, spineless, or unwilling to act. It would mean taking action, but doing so softly with the best interest of others at heart."[2]

I tried to imagine what that might look like in my life. Instead of being impatient about what I needed to get done while zipping up our boys' jackets, could I refocus my thoughts on smiling in appreciation for who they are? Could I do the same when I spoke to Peter? Could I put the best interests of others before my own to accept farm life in Minnesota as home? I knew I would need to change the tendency of my instinct to gratify myself before others. It was a nature that I'd fostered ever since I was little, ever since I competed with my sisters to get the most of a limited stack of pancakes and ever since I ventured, scared and alone, into the world at age twelve. While growing up, I'd often felt that if I didn't look out for me first, who would look out for me? I had had faith in God, but it was an immature faith. Now I was aware that God had brought me to Minnesota to grow my faith in him. Gentleness is a fruit of the spirit, something that God gives us when we let him dwell in us and guide our life. For me, fleeing my selfish instincts to pursue gentleness would require a deeper walk by faith.

"Will you forgive me, Peter?" I asked.

"I do," Peter said. Then he looked up at the clock. "Nine thirty! I better be on my way or I'll be gone all night." He leapt out of bed and back into his black Carhartts and sweatshirt.

"Don't you want to have intimacy?" I asked, wondering what I needed to do to fulfill his desires. "I'd be willing to do it for you."

"For us," Peter corrected me, looking straight into my eyes and smiling. He leapt back into bed.

As he held me in his arms, I thought how thankful I was for Peter's grace, a grace that was renewed each day. He loved me in spite of my shortcomings. Despite my selfishness, he had pursued me. This mirrored God's pursuit of me. Jesus left his perfect home in heaven to become human. He suffered the cross, taking the punishment I deserved and forgiving me, all because I was the joy set before him. He hadn't looked at my inner ugliness and turned away from me. Instead, he came after me. He called me by name. He pulled me out of darkness and into light. He washed me in his love because he had a purpose for me. All my life I'd thought that I needed to succeed in the world to become significant, but I was wrong. God's ways are not our ways. His word says, "Let no one deceive himself. If anyone among you seems to be wise in this age, let him become a fool that he may become wise. For the wisdom of this world is foolishness with God. For it is written, 'He catches the wise in their own craftiness'; and again, 'The LORD knows the thoughts of the wise, that they are futile.' Therefore let no one boast in men" (1 Cor. 3:18–21, NKJV). God hadn't made me to find my significance in nature or science. Nature and science were blessings and part of the life he had called me to, but the purpose of my life was to find God's love and be transformed by it. God intended for his love to change me, so that I would love others with his love, a self-sacrificing love. I knew that each day I would still fall short, but God would always forgive. His mercy would be new every morning. Each day he would give me his patience, gentleness, and love. All I had to do was ask for it because his faithfulness does not fail.

Peter left to finish the night's farm work. I turned out the light with his side of the bed empty. *Please bring him home tonight*, I prayed. Before falling asleep, I wondered what would become of me and the boys if something happened to Peter and he didn't come back. After all, farming is one of the deadliest professions in the world. The only reason Peter's dad took over our farm was that his brother, Peter's uncle, had died as a young father when the hydraulic support on a truck failed and the truck came down on him. He left a wife and three children to find their way in the world without him. This had taken place in the farmyard, only a hundred feet outside our bedroom window. I knew as I lay there that it would do no good to live or sleep in fear. God gives and he takes away. If something should ever happen to Peter, then I would still be in God's hands. *Nonetheless, God, if it is your will, please let me wake up and have Peter tomorrow*, I whispered before falling asleep.

...

THE NEXT MORNING, I ran fast along the green, grassy ditch trail heading south through the fields from our farm. The early morning air was crisp and cold. Gone were the walls of golden corn that had towered three feet above my head before the combine had passed through just days before. I was grateful for the fresh horizon, dotted by groves of hardwood trees amidst the never-ending open fields of harvest. I could see for miles in all directions now. I surprised myself when I admitted that I missed the sound of the wind rustling through the long, dry yellow leaves in the corn canopy. I missed looking up at tassels dancing like flowery yellow crowns against the pale blue autumn sky. I missed the little brown birds that would dart in and out of the tall corn grass in front of me, and I missed wondering what other animals were out there, crouching in the corn, watching silently as I ran by. I had known they were there because I had leapt over their burrows and seen the trails they had worn in the grass along my running path.

Soon I had to leap over a wide, black hose that ran along the trail beside me. The hose carried manure from far-away pig barns out onto the fields to fertilize crops. I resisted the urge to resent its ugliness and odor. I did not block out the bittersweet memories of running the pristine mountain trails overlooking glacial fjords in Cordova as they came to mind, but I also did not cling to my desire to hold on to the way things were in the past. Doing so would paralyze me from adapting to and appreciating the way things were in the present. It would prevent me from seeing the wondrous things around me and sharing a love for them with my family.

I looked beyond the manure hose and found gratitude for the water flowing in the ditch. I said so out loud, to God. I admired the ditch-side grasses and horsetails, which were home to a busy company of spiders, grasshoppers and dragonflies, frogs, red-winged blackbirds, a blue heron, field mice, a muskrat family, and a screech owl. The owl looked back at me and then took leave of the old ditch-side elm tree to hide in a distant island of woods, the woods I'd named Coyote Island. I made a mental note to bring my boys out that week with their field notebooks and magnifying glasses to explore and record.

I had to hurry now. The combine was in the last of our fields. Today was the final day of harvest, and I wanted to dress up pretty for a ride with Peter. Twelve years ago, when I was new to Norseland, I laughed when I heard someone say, "You know you're a farmwife when a date with your husband means riding the tractor with him." I'd sworn to myself that would not be me. I would rather wait until my husband was done with the tractor, and then we'd go on a proper date in town. But now, after years of waiting and weeks of seeing very little of him, the idea of sitting at his side sounded romantic.

Returning from my run, I packed Peter's lunch bucket and cut the kids' morning homeschool lessons short, setting them free to play outdoors in the sunshine that was steadily warming the day. If I was going to get in an hour of riding with Peter before my afternoon teleconferences, I'd need to hurry. By the time I reached the combine, the surprise was on me. I found Peter's cousin Eric was there to ride too. He'd taken time away from harvesting his own crops to get a good look at Peter's new German-built combine, a Claas Lexion 8600, which today did the work five hundred men would have been needed for to harvest corn by hand eighty years ago. I wanted to accept Eric's offer to leave so that Peter and I could ride alone, but then I realized that was wrong. The three of us crowded into the small cab together, me sitting on the lunch bucket in front of Peter's controls. We watched families of bunnies scurry out from the corn in front of the combine. The time had come for them to make a new home in the prairie grasses that skirted the fields. That is, if they could make it there before the coyotes lurking behind the harvest parade had their predatory fill.

After a round of combining across the field, Eric said goodbye. I crawled across Peter's lap to the open passenger seat and watched his thick, strong hands on the controls as he navigated the huge machine. With great care he steered around drainage tile inlets and electricity poles, all the while keeping pace with a large grain cart tractor that had come up alongside to receive the corn from the combine on the move. My husband was dusty and handsome. Sitting beside him, I felt as excited as if this were our first date, only now my love and admiration for this man were twelve years deep.

"You look nice." Peter winked, taking his eyes momentarily off the controls. He was quiet for a few minutes and then said, "This is the best yield we've had ever on this field. The green manure from cover crops seems to be improving the soil. Will you remind me next year to set up some experimental treatments here, to check this for sure?"

"Oh yes! That would be really exciting," I answered him.

That was it! I had found the answer to the question I had been asking in the tundra tent, about what I could possibly do to help my husband farm. It was research! I could get involved with Peter's farm research! I would try to do it with all my heart. After all, I was proud of my husband's hard work. I was thrilled that his years of planning, risk-taking, and implementing new and better soil management techniques were finally proving what he had long believed: what is good for the soil ecosystem is also good for the crop.

"Only a few more rounds to go," Peter said. "Won't that be a relief to have all of this year's crop in the bins!"

"I'll send the boys out to ride with you for the last of it," I said, climbing down out of the combine and setting off back to the farmhouse for my teleconference. It was particularly hard this day to sit indoors in front of the computer screen. I was glad when business was taken care of and I could swap my dress clothes for field garb. I tied a red bandana in my hair and rushed out to the garden to pull the remaining stalks of senesced basil, eggplant, tomatoes, and flowers left behind after Peter's mom had pulled what she could with her impressive eighty-three-year-old strength. I thought of all she had taught me about gardening and keeping house and about the example she set of unselfishly loving her husband. I worked in the garden until the sun hung low in the western sky. Its golden rays shown eastward, lighting up the giant green Claas combine a mile away heading at a steady clip toward home.

Like a statue in the garden, I froze. For twelve years I'd locked myself inside a prison of my own pride. This farm was a place I had resented. It had taken me away from the mountains, from the Arctic, and from Russia. It had taken me away from daily interactions with students and colleagues on campus and opportunities to pursue invitations for professorships at Harvard, UC Davis, Caltech, and the University of Waterloo in Canada. I had resolved to resent this farm when we got married. Pride had been my prison and resentment the bars through which I had looked out on the rural Midwest world. I knew that to become free I would have to continue to die to self. I would have to let my pride go in order to become something new, someone God was molding me into—something I trusted and at the same time feared would bring a joy that was not of myself.

One more layer of pride came loose and fell to the ground as I dropped my garden tools. I pulled the bandana from my head and raced the combine one-eighth of a mile to the driveway entrance to our farm. I wanted to give Peter and our boys a congratulatory welcome home. It was a tie! I jumped up and down in the driveway, waving my handkerchief and kicking my legs clumsily in the air in alternating directions as the huge machine turned into the property. Inside the cab, two boys and a man leaned forward in their seats, not believing their eyes but smiling and waving furiously at me as my heart filled up with joy and my tears overflowed. Since leaving home at twelve, I had felt no true home, until this moment, and I now realized that I would rather be nowhere other than in this family, in this place.

...

AS I STARTED to let go of my focus on self, withered relationships from my past began to bloom again. I looked out the kitchen window at Peter's hardworking father making his way to the barn, and wondered what a time-lapse camera would show of this farmyard as generation after generation had worn the same paths between buildings for 150 years. Perhaps, over the course of time, the skip and run in my boys' steps would slow to the stooped plodding their grandfather had now. I put the dishes down in the sink and ran outside with the compost to wish Peter's dad a good morning.

I spoke more often on the phone with my parents. My dad and Sandy, and at other times my mother, came out to our farm to visit us and spend time with the boys. I was gladdened to see the fun, loving relationships between them developing. My parents helped us with rock picking in the fields in spring and harvesting apples in the fall. My sisters with their husbands and children also came, sharing in the thrills of climbing in the hayloft, feeding cattle, and catching fireflies after dark. Even Sergey came to our farm, one wintry December day, to work with me on the idea for a new paper.

"Katya, now it is easier for me to come to you," he said. Upon entering the front door, he dropped an armload of Russian children's books on the floor among our shoes. Then, when he wasn't working with me on science in the barn that Peter had remodeled into a beautiful field laboratory for me, he spent hours each day reading Russian stories to our boys on the couch. He translated the stories in broken English, inserting the Russian words he thought they should know. I watched with a smile in my heart. But I wondered, why was he spending so much time reading to them? These were not his own grandchildren. Was he again teaching me an important lesson? Was he demonstrating to me an effective way of teaching my children Russian? Was he igniting in them a love for Russia and its culture and stories? Did he know how much this meant to me? Probably so. My boys still love to pull those books out and read them and ask me to read to them.

At supper one evening Sergey said there was another reason he'd come. He had long wanted to see the part of America he thought must still retain the character and strength that once had made the United States so great. Since I loved and admired him so, his praise for the Midwest helped me to appreciate it more myself. In the fall of 2019, Peter gave me the go-ahead to apply for a new Russian research grant. I, above all, was elated to learn that it was funded! Now we could make plans to go as a family to continue permafrost research at Sergey's science station.

Epilogue

The weaving of my roles as wife, mother, and scientist in an ever-changing world is growing smoother as I come to know the meaning of home as a place where one stays to tend the gardens of landscape and relationships. I realized the shift of mindset that required me to be thankful, despite the changes in my lifestyle, could also be helpful to my outlook on climate change in the world. As I look at the changes taking place around us now in the Arctic—the rapidly thawing permafrost; the losses of glaciers, ice sheets, and sea ice; the threat of polar bear and walrus habitat loss; the expansion of shrubs in the tundra; the intensification of wildfires; and the spread of disease in plants—there is a temptation to join the cry of some people to take a stand to prevent these changes on Earth. There is a tendency for all of us to want to keep things the way we remember them. There is a deep desire to keep the Arctic, in all of its beauty and splendor that Indigenous knowledge and Western science have revealed to us during the past 150 years, the same. But the Arctic changed to become how it is today, and change will continue to happen. We are well on our way to an Arctic that will be at least seven degrees warmer by the end of this century.[1] We can hope and strive to slow this warming,[2] but change itself is unpreventable. Change has always happened. If permafrost had not thawed in the past even before the impacts of human-induced climate warming, then the reindeer herders, scientists, and snowy owls would not be able to stand on pingos (ice-core hills in permafrost) looking out at vast wetland-filled drained lake basins in the tundra. It is thermodynamically impossible for change not to occur. Permafrost will continue to thaw, and we can expect faster and faster rates. Already in the past sixty years the area of thermokarst lakes near our home in Fairbanks has nearly doubled.[3] The newly formed lakes are spewing methane more than two times faster than old lakes. Carbon currently sequestered in permafrost peat of Siberian alases will not stay protected from the atmosphere forever. Once it thaws, this carbon will again enter into an active biogeochemical cycle, being converted from its current state of peat back into greenhouse gases by microbes or wildfires that will further warm the planet.[4]

What we do have a choice about is how we view and interact with these changes. We can let the idea of change become a festering wound inside us, or we can adapt to the changes, allowing them instead to become something we can perceive as opportunity for improvement. All of us should work to avoid wasting energy so that we can slow the rates of change, but I believe the biggest impacts on both our attitudes and the environment could be for more people to step outside and rekindle a long-lost love of the natural world. Not everybody has the means to travel to the Arctic or even to visit a national park, but everybody has the opportunity to turn off their electronic devices and take a walk outdoors through the woods, prairie, garden, or park near to where they live. They have the opportunity to crouch down and look into the heart of a flower to see what kinds of insects may be working inside, to lift up a rock and catch a glimpse of the magnificent, surprising organisms that rapidly dive into soil holes to evade the light. They have the opportunity to breath deep the scent of earth after a newly fallen rain, or to sit for a while on a bench listening to the wind rustling the leaves of the trees or to the different kinds of birds chirping in its branches, or to gaze at the sparkle of light in dewdrops on grass.

It is so much easier to be good stewards of the natural world around us if we get to know it. Knowing it and loving it comes from spending time with it. To take really good care of our children, spouse, parent, or friend we need to know them. We need to be willing to take care of their needs in the present. This means loving, teaching, and playing with our children today, not telling ourselves that we'll try to make time for them tomorrow. In the same way, we need to take care of our natural resources today. We can't just say that change is inevitable so we'll take care of it in the future. If Peter took this approach with farming, by not employing no-till and cover-crop options that preserve our soil today, then it would be only a matter of time before this valuable soil resource was gone. Our children and grandchildren would have less opportunity to cultivate food sustainably for their generations. It is important that each person fulfill their responsibilities in the present. Tomorrow will bring enough trouble of its own, but there is hope if today and tomorrow we each do our part to take the best care possible.

Silent night

Acknowledgments

I thank Elias Altman, my agent at Massie & McQuilkin, who sought me out with the invitation to write this, my first book. He taught me the basics of how to undertake such a project, particularly the proposal. Miles Doyle, at HarperCollins, believed enough in the book's potential that he read my proposal on his family vacation and expressed a winning desire to take the book on as a project. Lovely Anna Paustenbach became my primary and invaluable editor at HarperCollins. She gave me the courage to write the interiority of my story by making me feel understood and never judged. Her insightful questions and suggestions turned an amorphous block of writing into a sculpture.

The scientific research projects described in the book would not have been possible without the invaluable contributions of my students, postdocs, colleagues, and mentors: Chris Arp, Nancy Bigelow, Allen Bondurant, Donie Bret-Harte, Laura Brosius, Erin Carr, Jeff Chanton, Mimi Chapin, Terry Chapin, Joy Clein, Ronald Daanen, Anna Davydova, Sergei Davydov, Dmitri Draluk, Claude Duguay, Mary Edwards, Clayton Elder, Abraham Emond, Melanie Engram, Louise Farquharson, Jacques Finlay, Peter Frenzel, Steve Frolking, Marie Laure Geai, Sam Greene, Guido Grosse, Stas Gubin, Philip Hanke, Nick Hasson, Joanne Heslop, Martin Jeffries, Gary Jewison, Ben Jones, Miriam Jones, Mark Kessler, Charlie Koven, Josefine Lenz, Anna Liljedahl, Prajna Lindgren, Michelle Mack, Karla

Martinez-Cruz, Laurel McFadden, Franz Meyer, Charles Miller, Burke Minsley, Dan Nidzgorski, Ingmar Nitze, Laura Oxtoby, Casey Pape, Andy Parsekian, Chien-Lu Ping, Lawrence Plug, Vladimir Romanovsky, Caroline Ruppel, Torsten Sachs, Sudipta Sarkar, Hilary Saucy, Thomas Schneider von Deimling, Ted Schuur, Armando Sepulveda-Jáuregui, Janelle Sharp, Amy Strohm, Frederic Thalasso, Claire Treat, Natalie Tyler, Dragos Vas, Dave Verbyla, Alexandra Veremeeva, Qianlai Zhuang, Galya Zimova, Nikita Zimov, and Sergey Zimov.

Laura Oxtoby and Dragos Vas pulled me out of the Southcentral Alaska lake when I fell through the ice.

I owe thanks to my late step-grandfather, Sidney Robinson. Grandpa, as I called him, always had a sincere interest and time to find out what I was working on. He hoped for my future, paid my college application fees, and in his 80s, came with me one bitterly cold winter to Siberia. I owe my love for the natural world to my father and his brother, Eugene, who took me for walks as a child in the mountains and asked me stimulating questions about the things we saw. I thank my mother for her unconditional love, her prayers, and for sharing with me what she thinks is beautiful in difficult situations. I thank my sisters, Hanna, Christa, and Annie, for their feedback on the book and for their lifelong friendship and faithfulness despite the many miles that separate us. I am thankful to my parents-in-law, Willis and Rachel Anthony, for their grace toward me and for their good examples of marriage, ethics, and service.

The content of the book was improved by insightful feedback from Anna Reinemann and my brother-in-law, Dillon Sauer. I'm indebted to Ina Timling for producing artwork that not only helped clarify technical aspects of the writing but also added beauty.

Finally, I thank my husband, Peter, the love of my life, and our two boys, Jorgen and Anders, who were patient with me while I wrote. These three, by their very being, give me reason to smile every day.

Notes

Chapter 1: An American Girl in Cherskii

1. S. A. Zimov et al., "North Siberian Lakes: A Methane Source Fueled by Pleistocene Carbon," *Science* 277, no. 5327 (1997): 800–2.

2. Greenhouse gases in the atmosphere warm the planet by acting like a blanket, trapping the sun's heat. While methane is far less abundant in the atmosphere than carbon dioxide, it is a more potent greenhouse gas and accounts for roughly 25 percent of the radiative force driving climate change. The century-scale global warming potential of methane—GWP_{100}—is approximately 28 times greater than that of carbon dioxide; over twenty years, methane's GWP is 84.

   A. L. Ganesan et al., "Advancing Scientific Understanding of the Global Methane Budget in Support of the Paris Agreement," *Global Biogeochemical Cycles* 33 (2019): 1475–512; and G. Myhre et al., "Anthropogenic and Natural Radiative Forcing," in IPCC, *Climate Change 2013: The Physical Science Basis. Contribution of Working Group I to the Fifth Assessment Report of the Intergovernmental Panel on Climate Change*, ed. T. F. Stocker et al. (Cambridge, UK, and New York: Cambridge Univ. Press, 2013), 659–740, https://www.ipcc.ch/site/assets/uploads/2018/02/WG1AR5_Chapter08_FINAL.pdf.

3. For a description of the formation of thermokarst lakes in general, see L. J. Plug and J. J. West, "Thaw Lake Expansion in a Two-Dimensional Coupled Model of Heat Transfer, Thaw Subsidence, and Mass Movement," *Journal of Geophysical Research* 114, no. F1 (2009): F01002, 18, https://agupubs.onlinelibrary.wiley.com/doi/epdf/10.1029/2006JF000740.

4. Polar ice core records indicate that during ice ages, carbon dioxide concentration in the atmosphere was around 200 parts per million, and during the warmer interglacial periods, the levels were around 280 parts per million. Methane concentrations were lower, around 375 parts per billion during ice ages and 680 parts per billion during interglacials. Since the start of the Industrial Revolution, the concentration of carbon dioxide in the atmosphere has increased annually, and the

pace of increase is accelerating, leading to global warming. Methane has also mostly increased, although stabilization of concentrations between 2000 and 2007 followed by a continuation in its rise have scientists scratching their heads. In 2020, global atmospheric carbon dioxide measured at the National Oceanic and Atmospheric Administration (NOAA) remote sampling stations was 412.5 parts per million; global average methane concentration in December 2020 was 1,892.3 parts per billion, which is the highest so far on record.

IPCC, *Climate Change 2001: The Scientific Basis. Contribution of Working Group I to the Third Assessment Report of the Intergovernmental Panel on Climate Change*, ed. J. T. Houghton et al. (Cambridge, UK, and New York: Cambridge Univ. Press, 2001), 881, https://www.ipcc.ch/site/assets/uploads/2018/03/WGI_TAR_full_report.pdf; K. Kawamura et al., "Northern Hemisphere Forcing of Climatic Cycles in Antarctica over the Past 360,000 Years," *Nature* 448 (2007): 912–16; P. O. Hopcroft et al., "Understanding the Glacial Methane Cycle," *Nature Communications* 8, no. 14383 (2017), https://www.nature.com/articles/ncomms14383/; and NOAA, "Despite Pandemic Shutdowns, Carbon Dioxide and Methane Surged in 2020," *NOAA Research News*, April 7, 2021, https://research.noaa.gov/article/Art MID/587/ArticleID/2742/Despite-pandemic-shutdowns-carbon -dioxide-and-methane-surged-in-2020.

5. M. C. Serreze et al., "Observational Evidence of Recent Change in the Northern High-Latitude Environment," *Climatic Change* 46 (2000): 159–207; and M. Sturm, D. K. Perovich, and M. C. Serreze, "Meltdown in the North," *Scientific American* 289, no. 4 (2003): 60–67.

6. During the last Ice Age, when the tilt, wobble, and orbital pathway of Earth around the sun aligned to minimize the amount of solar radiation reaching the planet, wintertime snowfall did not entirely melt in summer. This caused much of North America and Europe to become covered by ice sheets and glaciers. In areas where precipitation was too low for glaciers to form, extreme cold winters and warm summers supported vast grasslands. These grasslands extended across great swaths of Alaska and much of North Siberia and Eurasia, covering almost fifteen million square miles of the unglaciated parts of the Northern Hemisphere—an area equal to the size of the North American continent.

    As the glaciers ground and pulverized rock, winds blew the resulting flour of the glacial mill, transporting it around the Northern

Hemisphere. This aeolian silt settled on the unglaciated, arid grass-lands of North Siberia and Alaska. Nutrients in the silt fertilized growth of grasses that supported the great grazing animals of the Ice Age: mammoths, lions, bison, horses, and musk oxen. By trampling, grazing, and depositing excrement, this mammalian megafauna in turn controlled the vegetation and soil moisture, giving advantage to grass-dominated steppe over a steppe dominated by mosses or shrubs.

The dust fell at such a rate, and for so long, that grasses themselves became buried in silt. Animals were buried in silt where they died, as evidenced from the plethora of bones and occasional intact carcasses found in excavations of these icy, permafrost soils known today by scientists as yedoma. Ice Age winters much colder than those we have today caused the ground to freeze and remain frozen below the sur-face, even in summer. In winter, the ground cracked. Snow meltwater ran into the cracks in spring and refroze in the ground, leading to the formation of massive, dark ice wedges in the ground. The dust piled on, year after year, freezing in place each winter and building the ice-supersaturated yedoma to depths of 30, 100, or even 200 feet. Formation of these perennial frozen yedoma permafrost soils locked away billions of tons of Ice Age carbon and precipitation in an icy, underground freezer.

S. A. Zimov et al., "Steppe-Tundra Transition: A Herbivore-Driven Biome Shift at the End of the Pleistocene," *American Naturalist* 146, no. 5 (1995): 765–94; R. D. Guthrie, "Origin and Causes of the Mam-moth Steppe: A Story of Cloud Cover, Woolly Mammal Tooth Pits, Buckles, and Inside-Out Beringia," *Quaternary Science Reviews* 20 (2001): 549–74; and J. B. Murton et al., "Paleoenvironmental Inter-pretation of Yedoma Silt (Ice Complex) Deposition as Cold-Climate Loess, Duvanny Yar, Northeast Siberia," *Permafrost and Periglacial Processes* 26, no. 3 (2015): 208–88.

7. S. A. Zimov et al., "Contribution of Disturbance to Increasing Sea-sonal Amplitude of Atmospheric CO_2," *Science* 284, no. 5422 (1999): 1973–76.

8. S. A. Zimov et al., "North Siberian Lakes: A Methane Source Fueled by Pleistocene Carbon," *Science* 277, no. 5327 (1997): 800–2.

9. The atmosphere currently contains more than 830 gigatons of carbon (GtC). Sergey's calculations suggested that 450 GtC were stored in Pleistocene-aged yedoma permafrost soils. Analysis of more extensive

data later revealed a smaller soil carbon pool size for the yedoma region (211 +160/–153 GtC); however, when the refrozen yedoma deposits beneath drained thermokarst lakes were factored in, the more extensive data yielded a yedoma region soil carbon pool size of 456 ± 45 GtC, very similar to Sergey's original regional estimate.

P. Ciais et al., "Carbon and Other Biogeochemical Cycles," in IPCC, *Climate Change 2013: The Physical Science Basis. Contribution of Working Group I to the Fifth Assessment Report of the Intergovernmental Panel on Climate Change*, ed. T. F. Stocker et al. (Cambridge, UK, and New York: Cambridge Univ. Press, 2013), 465–570, https://www.ipcc.ch/site/assets/uploads/2018/02/WG1AR5_Chapter06_FINAL.pdf; S. A. Zimov et al., "Permafrost Carbon: Stock and Decomposability of a Globally Significant Carbon Pool," *Geophysical Research Letters* 33, no. 20 (2006): L20502, https://agupubs.onlinelibrary.wiley.com/doi/full/10.1029/2006GL027484; J. Strauss et al., "The Deep Permafrost Carbon Pool of the Yedoma Region in Siberia and Alaska," *Geophysical Research Letters* 40, no. 23 (2013): 6165–70; and K. M. Walter Anthony et al., "A Shift of Thermokarst Lakes from Carbon Sources to Sinks During the Holocene Epoch," *Nature* 511, no. 7510 (2014): 452–69.

Chapter 9: A Scientific Author

1. K. M. Walter et al., "Methane Bubbling from Siberian Thaw Lakes as a Positive Feedback to Climate Warming," *Nature* 443, no. 7107 (2006): 71–75.

2. K. M. Walter et al., "Thermokarst Lakes as a Source of Atmospheric CH_4 During the Last Deglaciation," *Science* 318, no. 5850 (2007): 633–36.

3. J. D. McDowell, More Than a Carpenter (Carol Stream, IL: Tyndale House, 1977); and L. Strobel, The Case for Faith: A Journalist Investigates the Toughest Objections to Christianity (Grand Rapids, MI: Zondervan, 1998) Pp. 416.

4. C. S. Lewis, *Mere Christianity* (New York: Macmillan, 1960), 52.

Chapter 10: Truth Pursuit

1. K. M. Walter, L. C. Smith, and F. S. Chapin III, "Methane Bubbling from Northern Lakes: Present and Future Contributions to the Global

Methane Budget," *Philosophical Transactions of the Royal Society A: Mathematical, Physical and Engineering Sciences* 365, no. 1856 (2007): 1657–76.

2. Many publications on the topic of permafrost carbon have largely been archived through the Permafrost Carbon Network: http://www.perma frostcarbon.org/publications.html.

Chapter 11: Peat Cakes and Wedding Cakes

1. After combining measurements of methane emissions from present-day thermokarst lakes with paleo records of thermokarst-lake initiation events since the Last Glacial Maximum, my research concluded that thermokarst lakes contributed substantially to new sources of Northern Hemisphere methane during the last deglaciation. Additional data, synthesized by L. S. Brosius et al., demonstrate that widespread lake formation in the early Holocene was a response to climate warming at that time. Projected twenty-first-century global warming is two orders of magnitude faster (decadal warming) than deglacial warming (millennial warming), and the associated permafrost carbon emissions are anticipated to be up to 900 times greater.

K. M. Walter et al., "Thermokarst Lakes as a Source of Atmospheric CH_4 During the Last Deglaciation," *Science* 318, no. 5850 (2007): 633–36; L. S. Brosius et al., "Using the Deuterium Isotope Composition of Permafrost Melt Water to Constrain Thermokarst Lake Contributions to Atmospheric CH_4 During the Last Deglaciation," *Journal of Geophysical Research: Biogeosciences* 117, no. G1 (2012): G01022; L. Brosius et al., "Spatiotemporal Patterns of Northern Lake Formation Since the Last Glacial Maximum," *Quaternary Science Reviews* 253 (2021): 106773, 12; and K. M. Walter Anthony et al., "Methane Emissions Proportional to Permafrost Carbon Thawed in Arctic Lakes Since the 1950s," *Nature Geoscience* 9, no. 9 (2016): 679–82.

2. Sergey and Nikita Zimov had combed beaches along headwalls where permafrost was actively thawing to collect bones and fossilized teeth remains of mammoths, musk oxen, lions, rhinos, and other large mammals that lived in North Siberia during the last Ice Age. The bones would tell them the density of animals this region could support if it were a grassland, as it had been during the last Ice Age. This information would be used to resurrect an Ice Age biome in a nature

preserve south of Cherskii, which they called Pleistocene Park. Their aim was to repopulate Siberia with large herbivores to restore a highly productive grazing ecosystem, similar to that which dominated the region thousands of years ago before the animals were either hunted out by humans or suffered extinction due to a rapidly changing climate. The Zimovs were confident that restoration of these animals was a promising means to help fight climate change today. The animals would do their part to help keep the Siberian permafrost frozen: grazers would maintain a grassland ecosystem that (1) reflects incoming solar radiation, (2) stores carbon belowground, and (3) gets trampled in winter, allowing cold air temperatures to penetrate into the ground to keep permafrost cold. For more information about Pleistocene Park, see S. A. Zimov, "Pleistocene Park: Return of the Mammoth's Ecosystem," *Science* 308, no. 5723 (2005): 796–98; R. Anderson, "Welcome to Pleistocene Park," *The Atlantic*, April 2017, https://www.theatlantic .com/magazine/archive/2017/04/pleistocene-park/517779/; and the Pleistocene Park website, https://pleistocenepark.ru/.

3. Guido Grosse was involved with fieldwork and remote sensing studies of North Siberian permafrost characteristics that yielded articles in numerous publications dating as early as 2000. A few examples of his work include: G. Grosse et al., "Geological and Geomorphological Evolution of a Sedimentary Periglacial Landscape in Northeast Siberia During the Late Quaternary," *Geomorphology* 86 (2007): 25–51, https://permafrost.gi.alaska.edu/sites/default/files/Grosse%20et%20 al%202007%20Geomorphology.pdf; L. Schirrmeister et al., "Periglacial Landscape Evolution and Environmental Changes of Arctic Lowland Areas for the Last 60,000 Years (Western Laptev Sea Coast, Cape Mamontov Klyk)," *Polar Research* 27, no. 2 (2008): 249–72, https:// polarresearch.net/index.php/polar/article/view/2886/6513; and A. A. Andreev et al., "Weichselian and Holocene Palaeoenvironmental History of the Bol'shoy Lyakhovsky Island, New Siberian Archipelago, Arctic Siberia," *Boreas* 38, no. 1 (2009): 72–110.

Chapter 16: A Sleeping Giant

1. I. S. A. Isaksen et al., "Strong Atmospheric Chemistry Feedback to Climate Warming from Arctic Methane Emissions," *Global Biogeochemical Cycles* 25, no. 2 (2011): GB2002, https://atmos.uw.edu/academics /classes/2011Q2/558/IsaksenGB2011.pdf.

2. K. M. Walter Anthony et al., "Geologic Methane Seeps Along Boundaries of Arctic Permafrost Thaw and Melting Glaciers," *Nature Geoscience* 5 (2012): 419–26.

Chapter 17: Farm Wife

1. K. M. Walter Anthony et al., "A Shift of Thermokarst Lakes from Carbon Sources to Sinks During the Holocene Epoch," *Nature* 511 (2014): 452–56.

2. K. M. Walter Anthony et al., "Methane Emissions Proportional to Permafrost Carbon Thawed in Arctic Lakes Since the 1950s," *Nature Geoscience* 9, no. 9 (2016): 679–82; and K. M. Walter Anthony et al., "21st-Century Modeled Permafrost Carbon Emissions Accelerated by Abrupt Thaw Beneath Lakes," *Nature Communications* 9, no. 3262 (2018), https://www.nature.com/articles/s41467–018–05738–9#citeas.

3. K. Schaefer et al., "The Impact of the Permafrost Carbon Feedback on Global Climate," *Environmental Research Letters* 9 (2014): 085003; C. D. Koven, D. M. Lawrence, and W. J. Riley, "Permafrost Carbon—Climate Feedback Is Sensitive to Deep Soil Carbon Decomposability but Not Deep Soil Nitrogen Dynamics," *PNAS* 112, no. 12 (2015): 3752–57; C. D. Koven et al., "A Simplified, Data-Constrained Approach to Estimate the Permafrost Carbon–Climate Feedback," *Philosophical Transactions of the Royal Society A* 373, no. 2054 (2015): 20140423, https://royalsocietypublishing.org/doi/pdf/10.1098/rsta.2014.0423; E. J. Burke et al., "Quantifying Uncertainties of Permafrost Carbon-Climate Feedbacks," *Biogeosciences* 14, no. 12 (2017): 3051–66; and A. D. McGuire et al., "Dependence of the Evolution of Carbon Dynamics in the Northern Permafrost Region on the Trajectory of Climate Change," *PNAS* 115, no. 15 (2018): 3882–87, https://www.pnas.org/content/pnas/115/15/3882.full.pdf.

4. C. D. Arp et al., "Threshold Sensitivity of Shallow Arctic Lakes and Sublake Permafrost to Changing Winter Climate," *Geophysical Research Letters* 43, no. 12 (2016): 6358–65; M. Langer et al., "Rapid Degradation of Permafrost Underneath Waterbodies in Tundra Landscapes—Toward a Representation of Thermokarst in Land Surface Models," *Journal of Geophysical Research: Earth Surface* 121, no. 12 (2016): 2446–70; and P. Roy-Leveillee and C. R. Burn, "Near-Shore Talik Development Beneath Shallow Water in Expanding

Thermokarst Lakes, Old Crow Flats, Yukon," *Journal of Geophysical Research: Earth Surface* 122, no. 5 (2017): 1070–89.

5. T. Schneider von Deimling et al., "Observation-Based Modelling of Permafrost Carbon Fluxes with Accounting for Deep Carbon Deposits and Thermokarst Activity," *Biogeosciences* 12, no. 11 (2015): 3469–88; and Koven, Lawrence, and Riley, "Permafrost Carbon—Climate Feedback Is Sensitive," 3752–57.

6. Walter Anthony et al., "21st-Century Modeled Permafrost Carbon Emissions Accelerated by Abrupt Thaw Beneath Lakes," *Nature Communications* 9, no. 3262 (2018), https://www.nature.com/articles/s41467-018-05738-9#citeas.

7. S. Fuss et al., "Betting on Negative Emissions," *Nature Climate Change* 4, (2014): 850–53; M. Meredith et al., "Polar Regions," in *IPCC Special Report on the Ocean and Cryosphere in a Changing Climate,* ed. H.-O. Pörtner et al. (prepared 2019; in press), 203–320, https://www.ipcc.ch/site/assets/uploads/sites/3/2019/11/07_SROCC_Ch03_FINAL.pdf.

8. Fuss et al., "Betting on Negative Emissions," 850–53.

Chapter 18: Spying on Methane

1. K. M. Walter et al., "The Potential Use of Synthetic Aperture Radar for Estimating Methane Ebullition from Arctic Lakes," *Journal of the American Water Research Association* 44, no. 2 (2008): 305–15; M. Engram et al., "Synthetic Aperture Radar (SAR) Backscatter Response from Methane Ebullition Bubbles Trapped by Thermokarst Lake Ice," *Canadian Journal of Remote Sensing* 38, no. 6 (2013): 667–82; and M. Engram et al., "Remote Sensing Northern Lake Methane Ebullition," *Nature Climate Change* 10 (2020): 511–17.

Chapter 19: Love at Last

1. K. Martin, "Gentleness," in *In God's Orchard* (Milwaukee, WI: Northwestern Publishing , 2019), 70.

2. Martin, "Gentleness," in *In God's Orchard*, 71.

Epilogue

1. The terrestrial Arctic is projected to warm 4 to 6 degrees Celsius under RCP4.5 and more than 7 degrees Celsius following RCP8.5 this century.

A. D. McGuire et al., "Dependence of the Evolution of Carbon Dynamics in the Northern Permafrost Region on the Trajectory of Climate Change," *PNAS* 115, no. 15 (2018): 3882–87; IPCC, 2013: Summary for Policymakers. In: Climate Change 2013: The Physical Science Basis. Contribution of Working Group I to the Fifth Assessment Report of the Intergovernmental Panel on Climate Change [Stocker, T.F., D. Qin, G.-K. Plattner, M. Tignor, S.K. Allen, J. Boschung, A. Nauels, Y. Xia, V. Bex and P.M. Midgley (eds.)]. Cambridge University Press, Cambridge, United Kingdom and New York, NY, USA. (Fig. SPM.8,Page 22)

2. E. G. Nisbet et al., "Methane Mitigation: Methods to Reduce Emissions, on the Path to the Paris Agreement," *Reviews of Geophysics* 58, no. 1 (2020): e2019RG000675.

3. K. M. Walter Anthony et al., "Decadal-Scale Hotspot Methane Ebullition Within Lakes Following Abrupt Permafrost Thaw," *Environmental Research Letters* 16 (2021): 035010, https://iopscience.iop.org/article/10.1088/1748-9326/abc848/pdf.

4. K. M. Walter Anthony et al., "A Shift of Thermokarst Lakes from Carbon Sources to Sinks During the Holocene Epoch," *Nature* 511, no. 7510 (2014): 452–69.